AF361827

Advanced Mechanical Modeling of Nanomaterials and Nanostructures

Advanced Mechanical Modeling of Nanomaterials and Nanostructures

Editors

Francesco Tornabene
Rossana Dimitri

MDPI • Basel • Beijing • Wuhan • Barcelona • Belgrade • Manchester • Tokyo • Cluj • Tianjin

Editors
Francesco Tornabene
Engineering for Innovation
University of Salento
Lecce
Italy

Rossana Dimitri
Engineering for Innovation
University of Salento
Lecce
Italy

Editorial Office
MDPI
St. Alban-Anlage 66
4052 Basel, Switzerland

This is a reprint of articles from the Special Issue published online in the open access journal *Nanomaterials* (ISSN 2079-4991) (available at: www.mdpi.com/journal/nanomaterials/special_issues/mechanical_modeling).

For citation purposes, cite each article independently as indicated on the article page online and as indicated below:

LastName, A.A.; LastName, B.B.; LastName, C.C. Article Title. *Journal Name* **Year**, *Volume Number*, Page Range.

ISBN 978-3-0365-4916-3 (Hbk)
ISBN 978-3-0365-4915-6 (PDF)

Contents

About the Editors . . . . **vii**

Preface to "Advanced Mechanical Modeling of Nanomaterials and Nanostructures" . . . . . . **ix**

Rossana Dimitri and Francesco Tornabene
Special Issue on Advanced Mechanical Modeling of Nanomaterials and Nanostructures
Reprinted from: *Nanomaterials* **2022**, *12*, 2291, doi:10.3390/nano12132291 . . . . **1**

Masoumeh Soltani, Farzaneh Atoufi, Foudil Mohri, Rossana Dimitri and Francesco Tornabene
Nonlocal Analysis of the Flexural–Torsional Stability for FG Tapered Thin-Walled Beam-Columns
Reprinted from: *Nanomaterials* **2021**, *11*, 1936, doi:10.3390/nano11081936 . . . . **5**

Xiaohui Song, Shunli Wu and Rui Zhang
Computational Study on Surface Bonding Based on Nanocone Arrays
Reprinted from: *Nanomaterials* **2021**, *11*, 1369, doi:10.3390/nano11061369 . . . . **33**

Qijun Duan, Jun Xie, Guowei Xia, Chaoxuan Xiao, Xinyu Yang and Qing Xie et al.
Molecular Dynamics Simulation for the Effect of Fluorinated Graphene Oxide Layer Spacing on the Thermal and Mechanical Properties of Fluorinated Epoxy Resin
Reprinted from: *Nanomaterials* **2021**, *11*, 1344, doi:10.3390/nano11051344 . . . . **43**

Tamil Selvan Ramadoss, Yuya Ishii, Amutha Chinnappan, Marcelo H. Ang and Seeram Ramakrishna
Fabrication of Pressure Sensor Using Electrospinning Method for Robotic Tactile Sensing Application
Reprinted from: *Nanomaterials* **2021**, *11*, 1320, doi:10.3390/nano11051320 . . . . **57**

Raffaella Striani, Enrica Stasi, Antonella Giuri, Miriam Seiti, Eleonora Ferraris and Carola Esposito Corcione
Development of an Innovative and Green Method to Obtain Nanoparticles in Aqueous Solution from Carbon-Based Waste Ashes
Reprinted from: *Nanomaterials* **2021**, *11*, 577, doi:10.3390/nano11030577 . . . . **69**

Daniela A. Damasceno, R.K.N.D. Nimal Rajapakse and Euclides Mesquita
Atomistic Modelling of Size-Dependent Mechanical Properties and Fracture of Pristine and Defective Cove-Edged Graphene Nanoribbons
Reprinted from: *Nanomaterials* **2020**, *10*, 1422, doi:10.3390/nano10071422 . . . . **85**

Fumio Narita, Yinli Wang, Hiroki Kurita and Masashi Suzuki
Multi-Scale Analysis and Testing of Tensile Behavior in Polymers with Randomly Oriented and Agglomerated Cellulose Nanofibers
Reprinted from: *Nanomaterials* **2020**, *10*, 700, doi:10.3390/nano10040700 . . . . **101**

Abdullah H. Sofiyev, Francesco Tornabene, Rossana Dimitri and Nuri Kuruoglu
Buckling Behavior of FG-CNT Reinforced Composite Conical Shells Subjected to a Combined Loading
Reprinted from: *Nanomaterials* **2020**, *10*, 419, doi:10.3390/nano10030419 . . . . **115**

Subrat Kumar Jena, Snehashish Chakraverty and Francesco Tornabene
Buckling Behavior of Nanobeams Placed in Electromagnetic Field Using Shifted Chebyshev
Polynomials-Based Rayleigh-Ritz Method
Reprinted from: *Nanomaterials* **2019**, *9*, 1326, doi:10.3390/nano9091326 . . . . **131**

Yunlei Yin, Xinfei Zhao and Jie Xiong
Modeling Analysis of Silk Fibroin/Poly(-caprolactone) Nanofibrous Membrane under
Uniaxial Tension
Reprinted from: *Nanomaterials* **2019**, *9*, 1149, doi:10.3390/nano9081149 . . . . **147**

Bo Yang, Xianghe Peng, Sha Sun, Cheng Huang, Deqiang Yin and Xiang Chen et al.
Detwinning Mechanism for Nanotwinned Cubic Boron Nitride with Unprecedented Strength:
A First-Principles Study
Reprinted from: *Nanomaterials* **2019**, *9*, 1117, doi:10.3390/nano9081117 . . . . **159**

About the Editors

Francesco Tornabene

Francesco Tornabene is an Associate Professor at the University of Salento. He was born on 13 January 1978 in Bologna. In 2001, he received a National Patent Bologna (Italy) for the Industrial Invention: Friction Clutch for High Performance Vehicles Question BO2001A00442. He earned from the University of Bologna, Alma Mater Studiorum, an M.Sc. degree in Mechanical Engineering on 23/07/2003. In 2007, he earned a Ph.D. degree in Structural Mechanics at the University of Bologna. From 2011 to 2012, he became a Junior researcher; from 2012 to 2021, he was an Assistant Professor and Lecturer. For a long time, his scientific interests have been structural mechanics, solid mechanics, innovative and smart materials, computational mechanics and numerical techniques, damage and fracture mechanics. He is the author of more than 280 scientific publications, and collaborates with many national and international researchers and professors all over the world, as visible from his scientific publications. He is the author of 11 books and a member of the Editorial Boards for 44 international journals. He is also Editor-in-Chief for two international journals; furthermore, he is an Associate Editor for seven international journals. In recent years, he has received prestigious awards, e.g., "Highly Cited Researcher by Clarivate Analytics"(years 2018, 2019, and 2020), "Ambassador of Bologna Award"(2019), and "Member of the European Academy of Sciences"(since 2018). He collaborates as a reviewer with more than 260 prestigious international journals in the structural mechanics field. From 2012, his teaching activity has included dynamics of structures; computational mechanics; plates and shells; theory of structures; and structural mechanics. He is habilitated as a Full Professor in the area 08/B2 (structural mechanics).

Rossana Dimitri

Rossana Dimitri is an Associate Professor at the University of Salento (since 2019). She received from the University of Salento an M. Sc. degree in "Materials Engineering"in 2004, a Ph.D. degree in "Materials and Structural Engineering"in 2009, and a Ph.D. degree in "Industrial and Mechanical Engineering"in 2013. In 2005, she received from the University of Salento the "Best M. Sc. Thesis Price 2003-2004"; in 2013, she was awarded by the Italian Group for Computational Mechanics (GIMC) the 2013 ECCOMAS PhD Award. Her current interests include structural mechanics, solid mechanics, damage and fracture mechanics, contact mechanics, isogeometric analysis, high-performing computational methods, consulting in applied technologies and technology transfer. In 2010 and 2011, she received a research fellowship from the ENEA Research Centre of Brindisi (UTTMATB-COMP). In 2011 and 2012, she was a visiting scientist with a fellowship at the Institut für Kontinuumsmechanik Gottfried Wilhelm Leibniz Universität Hannover. From 2013 to 2016, she was a researcher at the University of Salento. From 2013 to 2019, she collaborated as researcher RTD-B with the University of Bologna and the Texas A&M University for a comparative assessment of some advanced numerical collocation methods with lower computational cost for fracturing problems and structural modelling of composite plates and shells, made by isotropic, orthotropic and anisotropic materials. She is the author of 112 scientific publications, and she collaborates with many national and international researchers and professors worldwide, as visible from her scientific publications. She also collaborates with different prestigious international journals in the structural mechanics field, as a reviewer, member of editorial boards, and guest editor for different special issues.

Preface to "Advanced Mechanical Modeling of Nanomaterials and Nanostructures"

The continuous development of novel composite materials with increased mechanical performances and a low density has encouraged the adoption of different components, such as functionally graded materials (FGMs), carbon nanotubes (CNTs), graphene nanoplatelets, and SMART constituents for many practical engineering applications, e.g., biomedicine, aerospace facilities, the automotive industry, energy devices, and civil applications. In a context where the recent requirements in design and manufacturing have led to an increased development of nanoshells, carbon nanotubes, and paramagnetic nanoparticles, this book aims to gather together experts and young researchers for the mechanical modeling of materials and structures at a small scale. The physical and mechanical properties of small-scale structures are well known to be size-dependent. This represents a key aspect largely explored both theoretically and computationally by means of advanced nonlocal approaches. These are explored here to address both continuum solid mechanics and fracture mechanics, for which the nonlocal aspect is a requisite for a realistic description of fracture, including the crack inception or propagation and the structural size effect related to the existence of a finite-size fracture process zone. Advanced theories and high-performance computational modeling on the statics or dynamics of nanosystems and nanostructures are welcome, together with the development of enhanced nonlocal damage and fracturing models, able to capture the formation and propagation of the internal cracks related to the heterogeneity of complex materials and interfaces.

Francesco Tornabene and Rossana Dimitri
Editors

 *nanomaterials*

Editorial

Special Issue on Advanced Mechanical Modeling of Nanomaterials and Nanostructures

Rossana Dimitri * and **Francesco Tornabene** *

Department of Innovation Engineering, Università del Salento, Via per Monteroni, 73100 Lecce, Italy
* Correspondence: rossana.dimitri@unisalento.it (R.D.); francesco.tornabene@unisalento.it (F.T.)

Citation: Dimitri, R.; Tornabene, F. Special Issue on Advanced Mechanical Modeling of Nanomaterials and Nanostructures. *Nanomaterials* **2022**, *12*, 2291. https://doi.org/10.3390/nano12132291

Received: 24 June 2022
Accepted: 29 June 2022
Published: 4 July 2022

Publisher's Note: MDPI stays neutral with regard to jurisdictional claims in published maps and institutional affiliations.

The increased requirements in design and manufacturing nanotechnology have favored the development of enhanced composite materials with tailored properties, such as functionally graded (FG) and carbon-based materials, primarily carbon nanotubes (CNTs), and graphene sheets or nanoplatelets, because of their remarkable mechanical properties, electrical conductivity, and high permeability. In such a context, nanoscaled structural elements such as nanobeams and nanoplates are being widely adopted as key components in different modern engineering devices, including sensors, actuators, nanoelectromechanical systems (NEMS), transistors, probes, among others. The complicated nature of similar structural systems requires a proper investigation of their fundamental properties, from an experimental, theoretical, and computational perspective. In line with the experimental findings, classical continuum theories are unable to interpret realistically the physical and mechanical properties of nanomaterials and nanostructures, whereas nonlocal formulations are more prone to explore their possible size-dependence in most static, dynamic, fracture mechanics problems.

This Special Issue has collected 11 papers on the application of high-performing computational strategies and enhanced theoretical formulations to solve a wide variety of linear or nonlinear problems in a multiphysical sense, together with different experimental evidences. Thus, classical and nonclassical theories have been proposed and compared together with multiscale approaches and homogenization techniques for different structural members and practical examples.

More specifically, the first paper authored by M. Soltani et al. [1] leverages a Vlasov thin-walled beam theory and energy method to study the nonlocal flexural-torsional stability of FG tapered thin-walled beam-columns, accounting for the coupled interaction among axial and bending forces. The differential quadrature method (DQM) is selected by the authors as an efficient numerical strategy to solve the coupled governing equations of the problem, checking for the sensitivity of the response for different input parameters, such as the power-law index, nonlocal parameter, axial load eccentricity, mode number and tapering ratio, as useful for design purposes. In the further work by A.H. Sofiyev et al. [2], a Donnel-type shell theory and first order shear deformation assumptions are applied to assess the buckling behavior of FG-CNT reinforced composite conical structures under a combined axial/lateral or axial/hydrostatic loading condition. The governing equations of the problem are transformed into algebraic equations using the Galerkin procedure, while providing an analytical expression for the critical value of the combined loading, as useful for different practical engineering systems and devices.

In the same setting, several works from the literature have recently focused on structures embedding FG materials, even with possible defects and porosities, and have developed innovative analytical and numerical models combined with different higher-order assumptions for many multiphysical problems [3–7] and multiscale electromechanical applications [8–10]. With the continuous application of carbon nanomaterials, modified graphene materials doped with epoxy resin have become a key topic for the design and development of novel epoxy composites. In such a context, the molecular dynamics (MD) simulation technology has been successfully applied in the work by Q. Duan et al. [11] for the evaluation of the effect of a fluorinated graphene oxide layer spacing on the thermo-mechanical properties of fluorinated epoxy resins (F-EP). In this paper, it is found that a fluorinated graphene oxide (FGO) with ordered filling can significantly improve the thermo-mechanical properties of F-EP, and the modification effect is better than that of a FGO with a disordered filling. The same computational tool is, moreover, applied by X. Song et al. [12] for the computational study of the surface bonding based on nanocone arrays, accounting for the effects of separation distance, contact length, temperature and size of cones. The main findings from [12] have revealed to be useful in designing advanced metallic bonding processes at low temperatures and pressure with tenable performances. In line with the previous two works, in the work authored by B. Yang et al. [13], the MD is applied to study the thermal stability of the nanotwinned diamond and synthesized nanotwinned cubic boron nitride for nanotwinned structures with enhanced mechanical properties.

Today, graphene nanoribbons (GNR)-based materials represent a valid alternative for the reinforcement of nanoelectronics structures, due to their outstanding properties. Their synthesis process, however, cannot avoid the presence of possible defects such as Stone–Wales (SW), extended line of defects (ELD), and nanopores that can compromise the integrity of structures. This is gradually leading to an increased amount of works focusing on the effect of defects and edges on the mechanical properties of GNR, primarily applying MD simulations [14–16]. Among coupled problems, the work by S.K. Jena et al. [17] provides a useful investigation on the buckling behavior of Euler–Bernoulli nanobeams immersed within an electro-magnetic field, based on the Eringen's nonlocal assumptions and Rayleigh-Ritz method. The authors propose shifted Chebyshev polynomials to avoid any ill-conditioning of the system for a higher number of terms in the approximation due to the orthogonality of functions.

Among the large number of materials suitable for practical applications in tissue engineering, silk fibroin (SF) obtained from Bombyx mori silkworms is largely applied as a scaffold material due to its excellent biocompatibility and low immune reaction [18,19]. As the main factors affecting the mechanical properties of fibrous membranes are the micro-structure and single fiber properties of fibrous membranes, in the work [20], the authors establish the uniaxial tensile force relationship between single fibers and fibrous membranes by means of a micro-mechanical approach. The applicability of the analytical model proposed in [20] is evaluated comparatively against the experimental predictions from the literature. A different multi-scale mechanics concept is applied in [21], combined to a classical finite element approach, to assess the tensile behavior in polymers with randomly oriented and agglomerated cellulose nanofibers, while exploring the property interactions across different length scales.

At the same time, the extensive use of piezoelectric-based pressure sensors for tactile sensing applications has increased the sensibility of the scientific community versus polystyrene in lieu of polymer/composite or crystal materials due to its low-cost fabrication method and large area fabrication. In this context, the work authored by T.S. Ramadoss et al. [22] proposes an inexpensive atactic polystyrene as base polymer and fabricate functional fibers, and an electrospinning method is used. The authors study experimentally the fiber morphologies by using a field-emission scanning electron microscope, while proposing a unique pressure sensor fabrication method. In the last Ref. [23], some possible industrial applications of nanomaterials obtained from the waste ashes are suggested, including, for example, inks for Aerosol Jet® Printing, with useful insights from a sustainable perspective.

Although this Special Issue has been closed, further developments on the theoretical and computational modeling of enhanced nanocomposite materials and structures are expected, including their static, dynamic, and buckling responses, as well as their fracture mechanics, and these will be useful for many industrial applications.

Conflicts of Interest: The authors declare no conflict of interest.

References

1. Soltani, M.; Atoufi, F.; Mohri, F.; Dimitri, R.; Tornabene, F. Nonlocal Analysis of the Flexural–Torsional Stability for FG Tapered Thin-Walled Beam-Columns. *Nanomaterials* **2021**, *11*, 1936. [CrossRef] [PubMed]
2. Sofiyev, A.H.; Tornabene, F.; Dimitri, R.; Kuruoglu, N. Buckling behavior of FG-CNT reinforced composite conical shells subjected to a combined loading. *Nanomaterials* **2020**, *10*, 419. [CrossRef] [PubMed]
3. Chen, D.; Yang, J.; Kitipornchai, S. Buckling and bending analyses of a novel functionally graded porous plate using Chebyshev-Ritz method. *Arch. Civ. Mech. Eng.* **2019**, *19*, 157–170. [CrossRef]
4. Gao, K.; Huang, Q.; Kitipornchai, S.; Yang, J. Nonlinear dynamic buckling of functionally graded porous beams. *Mech. Adv. Mater. Struct.* **2019**, *28*, 418–429. [CrossRef]
5. Allahkarami, F.; Tohidi, H.; Dimitri, R.; Tornabene, R. Dynamic Stability of Bi-Directional Functionally Graded Porous Cylindrical Shells Embedded in an Elastic Foundation. *Appl. Sci.* **2020**, *10*, 1345. [CrossRef]
6. Kiarasi, F.; Babaei, M.; Asemi, K.; Dimitri, R.; Tornabene, F. Three-Dimensional Buckling Analysis of Functionally Graded Saturated Porous Rectangular Plates under Combined Loading Conditions. *Appl. Sci.* **2021**, *11*, 10434. [CrossRef]
7. Dastjerdi, S.; Malikan, M.; Dimitri, R.; Tornabene, F. Nonlocal elasticity analysis of moderately thick porous functionally graded plates in a hygro-thermal environment. *Compos. Struct.* **2021**, *255*, 112925. [CrossRef]
8. Bodaghi, M.; Damanpack, A.R.; Aghdam, M.M.; Shakeri, M. Geometrically non-linear transient thermo-elastic response of FG beams integrated with a pair of FG piezoelectric sensors. *Compos. Struct.* **2014**, *107*, 48–59. [CrossRef]
9. Kulikov, G.M.; Plotnikova, S.V. An analytical approach to three-dimensional coupled thermoelectroelastic analysis of functionally-graded piezoelectric plates. *J. Intell. Mater. Syst. Struct.* **2017**, *28*, 435–450. [CrossRef]
10. Arefi, M.; Bidgoli, E.M.R.; Dimitri, R.; Bacciocchi, M.; Tornabene, F. Application of sinusoidal shear deformation theory and physical neutral surface to analysis of functionally graded piezoelectric plate. *Compos. Part B Eng.* **2018**, *151*, 35–50. [CrossRef]
11. Qijun, D.; Jun, X.; Guowei, X.; Chaoxuan, X.; Xinyu, Y.; Qing, X.; Zhengyong, H. Molecular dynamics simulation for the effect of fluorinated graphene oxide layer spacing on the thermal and mechanical properties of fluorinated epoxy resin. *Nanomaterials* **2021**, *11*, 1344. [CrossRef]
12. Song, X.; Wu, S.; Zhang, R. Computational study on surface bonding based on nanocone arrays. *Nanomaterials* **2021**, *11*, 1369. [CrossRef] [PubMed]
13. Yang, B.; Peng, X.; Sun, S.; Huang, C.; Yin, D.; Chen, X.; Fu, T. Detwinning mechanism for nanotwinned cubic boron nitride with unprecedented strength: A first-principles study. *Nanomaterials* **2019**, *9*, 1117. [CrossRef] [PubMed]
14. Fu, Y.; Ragab, T.; Basaran, C. The effect of Stone-Wales defects on the mechanical behavior of graphene nano-ribbons. *Comput. Mater. Sci.* **2016**, *124*, 142–150. [CrossRef]
15. Zhang, J.; Ragab, T.; Basaran, C. Comparison of fracture behavior of defective armchair and zigzag graphene nanoribbons. *Int. J. Damage Mech.* **2019**, *28*, 325–345. [CrossRef]
16. Damasceno, D.A.; Rajapakse, R.K.N.D.N.; Mesquita, E. Atomistic modelling of size-dependent mechanical properties and fracture of pristine and defective cove-edged graphene nanoribbons. *Nanomaterials* **2020**, *10*, 1422. [CrossRef]
17. Jena, S.K.; Chakraverty, S.; Tornabene, F. Buckling behavior of nanobeams placed in electromagnetic field using shifted Chebyshev polynomials-based Rayleigh-Ritz method. *Nanomaterials* **2019**, *9*, 1326. [CrossRef]
18. Cho, H.J.; Ki, C.S.; Oh, H.; Lee, K.H.; Um, I.C. Molecular weight distribution and solution properties of silk fibroins with different dissolution conditions. *Int. J. Biol. Macromol.* **2012**, *51*, 336–341. [CrossRef]
19. Park, S.Y.; Ki, C.-S.; Park, Y.H.; Jung, H.M. Electrospun Silk Fibroin Scaffolds with Macropores for Bone Regeneration: An In Vitro and In Vivo Study. *Tissue Eng. Part A* **2010**, *16*, 1271–1279. [CrossRef]
20. Yin, Y.; Zhao, X.; Xiong, J. Modeling analysis of silk fibroin/poly(ε-caprolactone) nanofibrous membrane under uniaxial tension. *Nanomaterials* **2019**, *9*, 1149. [CrossRef]
21. Narita, F.; Wang, Y.; Kurita, H.; Suzuki, M. Multi-Scale Analysis and Testing of Tensile Behavior in Polymers with Randomly Oriented and Agglomerated Cellulose Nanofibers. *Nanomaterials* **2020**, *10*, 700. [CrossRef] [PubMed]
22. Ramadoss, T.S.; Ishii, Y.; Chinnappan, A.; Ang, M.H.; Ramakrishna, S. Fabrication of pressure sensor using electrospinning method for robotic tactile sensing application. *Nanomaterials* **2021**, *11*, 1320. [CrossRef] [PubMed]
23. Striani, R.; Stasi, E.; Giuri, A.; Seiti, M.; Ferraris, E.; Corcione, C.E. Development of an innovative and green method to obtain nanoparticles in aqueous solution from carbon-based waste ashes. *Nanomaterials* **2021**, *11*, 577. [CrossRef] [PubMed]

 *nanomaterials*

Article

Nonlocal Analysis of the Flexural–Torsional Stability for FG Tapered Thin-Walled Beam-Columns

Masoumeh Soltani [1,*], Farzaneh Atoufi [1], Foudil Mohri [2], Rossana Dimitri [3,*] and Francesco Tornabene [3]

[1] Department of Civil Engineering, Faculty of Engineering, University of Kashan, Kashan 8731753153, Iran; atoufi1373@gmail.com

[2] Université de Lorraine, CNRS, Arts et Métiers ParisTech, LEM3, LabEx DAMAS, F-57000 Metz, France; foudil.mohri@univ-lorraine.fr

[3] Department of Innovation Engineering, University of Salento, 73100 Lecce, Italy; francesco.tornabene@unisalento.it

* Correspondence: msoltani@kashanu.ac.ir (M.S.); rossana.dimitri@unisalento.it (R.D.)

Abstract: This paper addresses the flexural–torsional stability of functionally graded (FG) nonlocal thin-walled beam-columns with a tapered I-section. The material composition is assumed to vary continuously in the longitudinal direction based on a power-law distribution. Possible small-scale effects are included within the formulation according to the Eringen nonlocal elasticity assumptions. The stability equations of the problem and the associated boundary conditions are derived based on the Vlasov thin-walled beam theory and energy method, accounting for the coupled interaction between axial and bending forces. The coupled equilibrium equations are solved numerically by means of the differential quadrature method (DQM) to determine the flexural–torsional buckling loads associated to the selected structural system. A parametric study is performed to check for the influence of some meaningful input parameters, such as the power-law index, the nonlocal parameter, the axial load eccentricity, the mode number and the tapering ratio, on the flexural–torsional buckling load of tapered thin-walled FG nanobeam-columns, whose results could be used as valid benchmarks for further computational validations of similar nanosystems.

Keywords: axially functionally graded materials; differential quadrature method; flexural–torsional buckling; nonlocal elasticity theory; tapered I-beam

Citation: Soltani, M.; Atoufi, F.; Mohri, F.; Dimitri, R.; Tornabene, F. Nonlocal Analysis of the Flexural–Torsional Stability for FG Tapered Thin-Walled Beam-Columns. *Nanomaterials* **2021**, *11*, 1936. https://doi.org/10.3390/nano11081936

Academic Editor: Francisco Torrens

Received: 31 May 2021
Accepted: 21 July 2021
Published: 27 July 2021

Publisher's Note: MDPI stays neutral with regard to jurisdictional claims in published maps and institutional affiliations.

1. Introduction

Thin-walled beams with open cross-sections (e.g., channel, angle, I- and Tee-sections) carry an extensive variety of potential applications as structural components in various engineering fields (from civil to aeronautical engineering) since they offer high performances with a minimal weight. Moreover, thin-walled beams with varying cross-sections have been of great interest to designers and researches, especially in recent decades. The optimization of weight, the reduction in volume, and the improvement of both strength and stability represent some crucial reasons to increase their use as structural members. Due to the low torsion stiffness, a slender beam with a thin-walled cross-section subjected to an eccentric compressive axial force can buckle in the flexural–torsional mode. Thus, investigations about the stability of tapered thin-walled beams can be very complicated because of the coupled bending and torsional deformations involved, as well as the arbitrary variation in the geometrical properties along the longitudinal direction.

As far as advanced multi-phase composites are concerned, functionally graded materials (FGMs) represent a novel generation of composite materials, based on a smooth and gradual variation in the volume fraction of their constituent phases in any desired direction. Compared to traditional materials and laminated composites, FGMs possess some important advantages, primarily, multifunctionality, a high temperature-withstanding ability, the reduction or total removal of stress concentrations, together with the improved strength and fracture toughness. Due to these favorable features, FGMs can represent ideal materials for the design of smart engineering systems and devices, which has motivated their recent extensive use in many engineering applications and modern industries, such as aerospace, automobile, optics, nuclear, electronic and turbine components.

With the recent development of nanotechnology, nanoscaled structural elements such as nanobeams and nanoplates are being widely used as key components in different modern engineering devices, including sensors, actuators, transistors, probes, and nano-electromechanical systems (NEMS). This requires an appropriate study of the mechanical properties of similar structural systems, with even more complicated natures. The experimental tests demonstrate that classical continuum theories cannot be implemented for the exact analysis of nanostructures, as the size effect can play a significant role in their mechanical behavior. Thus, various higher-order size-dependent continuum theories, such as the modified couple stress theory [1], the surface energy theory [2] and nonlocal elasticity theory [3,4], have been expanded to model small-sized structures. Among these models, the nonlocal elasticity theory, as suggested by Eringen [3], has been widely used in the literature to investigate the stability, deformation and vibrational responses of nanostructural elements, assuming that the stress state at an arbitrary point in a body depends not only on the strain field at that point, but also on the strain fields at all points of the body. At the same time, FGMs have been increasingly applied in small-sized structures due to their superior mechanical properties. In such a context, over the past few years, several investigations have been performed to study the linear and nonlinear mechanical responses of nanosized structures made from homogenous or FGMs. Moreover, a large number of works can be found in the literature focusing on the elastic and/or inelastic static, vibration and instability behavior of beams with a thin-walled cross-section, due to their vast relevance in many engineering configurations. Among the most relevant works on the topic, Kitipornchai and Trahair [5] determined the flexural–torsional critical force of doubly- and/or singly-symmetric I-beams with a geometrical variation under non-uniform torsion. Wekezer [6,7] studied the stability of thin-walled beams with varying open sections based on shell theory strain tensors. Considering the influence of geometric nonlinearity, a finite element technique was suggested by Yang and Yau [8] to assess the buckling behavior of doubly symmetric tapered I-beams. Bradford and Cuk [9] adopted a novel finite element technique to determine the buckling limit state of web-tapered beams with a mono-symmetric I-section. In another study, web-tapered beams with a Tee-section were probed by Baker [10]. A finite element formulation was also applied by Rajasekaran [11,12] to approximate the linear stability resistance of tapered thin-walled beams. Similarly, a simple finite element solution was presented by Gupta et al. [13] and Ronagh et al. [14] to predict the lateral–torsional resistance of tapered I-beams. With the help of the total potential energy and Hamilton's principle, Chen [15] computed the vibrational properties of thin-walled beams with geometrical variation. An innovative finite element formulation was also proposed by Kim [16] to analyze the lateral–torsional buckling (LTB) and vibration behavior of beams with a tapered I-section under different boundary conditions. The shear deformation effect was also accounted within the formulation of Li [17] for the stability study of beams with a linearly variable cross-section under a compressive axial load. A nonlocal elasticity version was also suggested by Peddieson et al. [18] to elaborate a nonlocal Benoulli/Euler beam model. A semi-inverse approach was then employed by Elishakoff et al. [19] for the vibrational analysis of beams made of axially-inhomogeneous materials, whereas Refs. [20–22] represent some further useful contributions to the lateral–torsional stability study of thin-walled beams with doubly- and singly-symmetric I-sections under different boundary conditions. Taking into account small deformations and large displacements, Mohri et al. [23,24] analyzed the nonlinear flexural–torsional behavior of thin-walled beams with arbitrary cross-sections by employing the Galerkin method, while Samanta and Kumar [25] provided a shell finite element solution for the study of the distortional buckling resistance of beams with a singly-symmetric I-section under simply supports.

In the field of nonlocal differential elasticity methodology, Reddy [26] proposed some pioneering analytical solutions for the static, buckling and vibrational analyses of beams by considering different shear deformation theories. Some additional analytical outcomes for cantilever beams with linear tapered section were also presented by Challamel et al. [27]. Wang et al. [28] perused the flexural vibration problem of nano- and microbeams, following the assumptions of the nonlocal elasticity theory of Eringen in conjunction with the Timoshenko beam model. Many further works in the literature have successfully applied the Eringen nonlocal elasticity approach combined with different beam theories and numerical solution methods—see Refs. [29–38]. Among them, Pradhan and Sarkar [29] studied the deformation, instability and vibrational responses of an Euler–Bernoulli beam with variable geometrical and material properties. Aydogdu [30] derived a generalized nonlocal beam theory for the mechanical analysis of nanosize beams by means of the Eringen elasticity assumptions combined with different beam theories. In the same direction, Civalek and Akgöz [31] studied the free vibrational properties of microtubules, which problem was solved numerically based on the DQM. Danesh et al. [32] determined the equations of motion for the longitudinal vibration of nanorods with tapered cross-sections, and solved them via the DQM. Şimşek and Yurtcu [33] used the Timoshenko beam theory to survey the deformation and buckling capacity of nanobeams with varying materials. McCann et al. [34] studied the lateral buckling resistance of steel beam members under pure bending and with simply-supported ends, in presence of discrete elastic lateral restraints along their axial direction. Following the nonlocal continuum theoretical assumptions, a finite element formulation was suggested by Eltaher et al. [35,36] to assess the size effect on the mechanical response of nanobeams made of FG materials. An Euler–Bernoulli beam model was also proposed by Shahba et al. [37] to compute the critical axial forces and natural frequencies of tapered beams with axially non-homogeneous materials. Within the framework of large torsion, Benyamina et al. [38] developed a nonlinear formulation to analyze the lateral stability and buckling moment of tapered I-section beams under simply–simply supports. Among the different numerical strategies to handle similar problems, Nguyen et al. [39] proposed an approximate methodology to evaluate the critical moment of I-section beams in the presence of discrete torsional bracing. Attard and Kim [40] included the shear deformations to determine the lateral stability equations for isotropic beams with a thin-walled open section. Challamel and Wang [41] employed Bessel functions for an exact computation of the lateral–torsional buckling load of strip cantilever beam members subjected to an arbitrary loading distribution. A modified couple stress theory was differently combined with the first-order shear deformable beam model of Ke et al. [42] to describe the size effect on the dynamic stability of microbeams made of FGMs. A novel finite element solution was proposed by Borbon [43] to study the coupled vibrational responses of beams with non-symmetric thin-walled cross-sections, accounting for the possible influence of loading eccentricities, shear deformation and rotatory inertia, and a further approximate methodology was successfully introduced by Serna et al. [44] to study the elastic flexural buckling of non-uniform columns subjected to arbitrary axial forces. Akgoz and Civalek [45] surveyed the free vibrational problem of axially functionally graded (AFG) non-uniform microbeams based on a Euler–Bernoulli beam model and modified couple stress theory. In order to exhaustively assess the static and dynamic responses of beams made of FG piezoelectric materials, an improved three-noded beam element was formulated by Lezgy-Nazargah et al. [46], whereas a novel beam finite element was developed by Trahair [47] for the lateral stability analysis of cantilever tapered steel beams. Different examples of nonlocal models and numerical methods can be found in the literature for a large variety of coupled problems and engineering applications. In Refs. [48,49], the authors proposed a Timoshenko beam nonlocal model to assess the free vibrational response of magneto-electro-elastic nanobeams [48], also made of FGMs [49]. The von Kármán geometric nonlinearity was included within a first-order shear deformable beam model by Liu et al. [50] in a nonlocal elasticity context, to evaluate the buckling and post-buckling responses of nanobeams made of piezoelectric materials in thermo-electro-mechanical conditions. A third-order shear deformable beam theory was adopted by Nami et al. [51] for a thermal stability analysis of FG nanoplates. Among tapered member applications, a novel beam finite element was introduced by Mohri et al. [52], together with a large torsion assumption, to estimate the stability resistance of tapered thin-walled beams. A semi-analytical procedure based on the Ritz technique was employed by Kuś [53] for analyzing the lateral stability of linearly tapered-web and/or flange doubly-symmetric I-beams. A finite element-based solution was proposed by Pandeya and Singhb [54] to survey the free vibrational behavior of a fixed–free nanobeam with a varying cross-section. According to the Eringen nonlocal theory and Euler–Bernoulli beam model, the nonlinear vibration of AFG nanobeams with a tapered section was exploited by Shafiei et al. [55], and a semi-analytical finite strip procedure was implemented by Zhang et al. [56,57] for the study of the stability capacity of bars with an open and closed cross-section under an axial loading condition [56], accounting for the effect of lateral elastic braces on the overall stability response in Ref. [57]. Further studies on the nonlocal vibration, buckling, and post-buckling of size-dependent beams, rods and plates at different scales can be found in [58–67], both in an analytical and a numerical sense. More specifically, as far as thin-walled structures are concerned, novel efficient models and computational methods have been developed in the literature to treat even more complicated applications. Among the most recent works, a novel optimization methodology was proposed by Maalawi [68] to enhance the vibrational response of thin-walled box beams with varying material properties. An innovative finite element formulation was also suggested by Lezgy-Nazargah [69] based on the theory of a generalized layered global–local beam (GLGB), to carry out an elasto-plastic analysis of thin-walled beams with reduced computational effort. Nguyen et al. [70,71] derived an efficient finite element formulation to investigate the flexural–torsional stability and buckling response of FGM beams with a singly symmetric open section, in the framework of Vlasov's theory. Li et al. [72] applied the method of generalized differential quadrature to rigorously solve the bending, buckling and vibrational problems of AFG beams, accounting for nonlocal strain gradient theoretical assumptions. Moreover, Khaniki et al. [73–80] published several important contributions elated to the static, vibrational and buckling analysis of small-size beams with a constant or variable cross-section, made of homogenous and/or FGMs. A finite element approach was recently developed by Koutoati [81] to assess the static and free vibrations of multilayer composites and FG beams by means of different shear deformation beam theories. Following the first-order shear deformation theory, Glabisz et al. [82] formulated an innovate algorithm to analyze the stability and vibrational problem of nanobeams incorporating different end supports. Within a modified shear deformation theory context, in which it is not essential to use the shear correction factor, the stability and free vibration behavior of FG nanobeams were explored by Ebrahim et al. [83] using the Chebyshev–Ritz method. A double analytical and finite element solution has recently been proposed by Jrad et al. [84] to assess the triply coupled free vibrational responses of thin-walled beams under different boundary conditions. More recently, a third-order shear deformation theory was employed by Arefi and Civalek [85] to check for the static deformation of cylindrical nanoshells made from FG piezoelectric materials supported by a Pasternak elastic foundation.

Among the studies on tapered structures, Osmani and Meftah [86] studied the shear deformation effect on the buckling response of tapered I-shape beams under different loading conditions. An innovative methodology based on the classical energy approach was expounded by Chen et al. [87] for predicting the lateral buckling resistance of I-beams with simple supports. Achref et al. [88] analytically assessed the higher-order instability loads of beams with thin-walled open cross-sections under different loading conditions by resorting to a classical finite element approach for comparative purposes. Different numerical approaches were applied in Refs. [89–93] for the linear stability and free vibrational study of homogenous and AFG tapered thin-walled beams with an open cross-section, subjected to different boundary conditions and arbitrary loading cases.

Based on the available literature, however, it seems that the flexural–torsional stability of AFG nanobeam-columns with tapered I-section has never been assessed. The current research is moving in this direction, and is aimed at probing the size-dependent buckling properties of AFG tapered nanobeams with a doubly-symmetric thin-walled cross-section, according to Vlasov assumptions. All the mechanical properties in the present work are graded in the longitudinal direction using the power function except, for the Poisson's ratio, wherein the small size effect is taken into account via the Eringen nonlocal elasticity theory. The nonlocal governing equations of the problem, together with the associated boundary conditions, are obtained by implementing the Vlosov model and the energy method, in order to account for the eccentricity effect of of a compressive axial loading from the centroid within the formulation. The DQM is here employed to solve the resulting stability equations in a strong form and to determine the flexural–torsional buckling load. Different numerical examples analyze the effects of several parameters, namely, the constituent volume fractions, tapering ratio, nonlocal parameter and mode number, on the flexural–torsional stability of AFG tapered nanobeams with an I-section subjected to simply supported boundary conditions. The work is organized as follows. After a preliminary description of the theoretical formulation (Section 2), we provide (in Section 3) the basic notions of the DQM, here applied as an efficient tool to solve the problem with reduced computational effort. In Section 4 we present the results from a large parametric investigation aimed at checking the sensitivity of the mechanical response to different input parameters, which is useful for design purposes. The main results and concluding remarks are discussed in Section 5.

2. Problem Definition

The following stability model represents an extension of the formulation proposed in Ref. [94] for non-prismatic thin-walled nanobeam-columns with an arbitrary distribution of the material properties in the axial direction, whose numerical outcomes could be useful for the development and design of thin-walled structures, such as scanning tunneling microscopes with nonuniform nanobeams at tunneling tips. Due to the rapid development of nanoscience, the stability of FG nanobeams with variable thin-walled cross sections represents one of their key design benefits, as here explored theoretically via nonconventional Eringen nonlocal elasticity, and numerically via the DQM.

2.1. Kinematics

Consider a straight tapered doubly symmetric I-beam made of non-homogeneous material, with variable properties along its longitudinal direction, as represented in Figure 1.

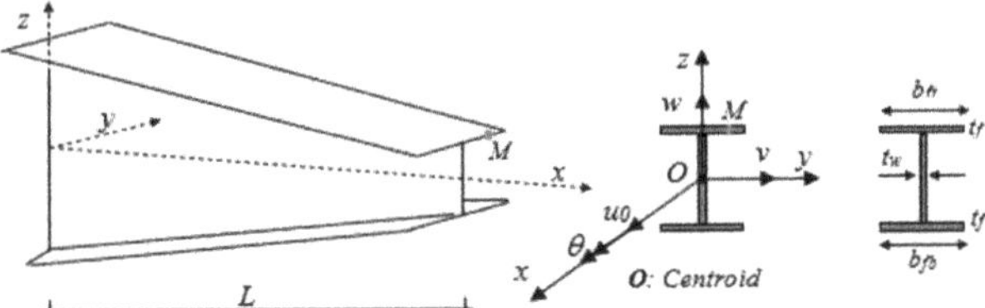

Figure 1. Geometrical scheme of variable doubly symmetric I-section beam—coordinate system and notation for the displacement parameters.

The orthogonal right-hand Cartesian coordinate system (x, y, z) is adopted, wherein x denotes the longitudinal axis, and y and z are the first and second principal bending axes parallel to the flanges and web, respectively. The origin O of these axes is located at the centroid of the cross-section. In the current work, it is assumed that the height of the web and/or width of both flanges can vary linearly along the longitudinal direction (x-axis), while the thickness remains constant. In the case of doubly-symmetric thin-walled sections, the shear center coincides with the centroid. In this study, we consider only slender beams, such that shear deformations can be ignored in our formulation, together with the local and distortional deformations. Based on these assumptions and following the Vlasov model for non-uniform torsion [95], the displacement field for an arbitrary point on the beam can be expressed as

$$U(x,y,z) = u(x) - y\frac{\partial v(x)}{\partial x} - z\frac{\partial w(x)}{\partial x} - \omega(y,z)\frac{\partial \theta(x)}{\partial x} \tag{1a}$$

$$V(x,y,z) = v(x) - z\theta(x) \tag{1b}$$

$$W(x,y,z) = w(x) + y\theta(x) \tag{1c}$$

In these equations, U is the axial displacement, V and W represent the lateral and vertical displacements (along the y- and z-directions, respectively); u,v,w are the kinematic quantities defined at the reference surface; $\omega(y,z)$ stands for the warping function for the variable cross-section, which can be defined based on St. Venant torsional theory, and θ is the twisting angle. The Green strain tensor components in the large displacement include both the linear and the nonlinear strain parts, as follows

$$\varepsilon_{ij} = \frac{1}{2}\left(\frac{\partial U_i}{\partial x_j} + \frac{\partial U_j}{\partial x_i}\right) + \frac{1}{2}\left(\frac{\partial U_k}{\partial x_i}\frac{\partial U_k}{\partial x_j}\right) = \varepsilon^l_{ij} + \varepsilon^*_{ij} \quad i,j,k = x,y,z \tag{2}$$

where ε^l_{ij} denotes the linear part, and ε^*_{ij} refers to the quadratic nonlinear part. For thin-walled beams, the strain tensor components reduce to the following:

$$\varepsilon_{xx} \approx U' + \frac{1}{2}\left(V'^2 + W'^2\right) = \varepsilon^l_{xx} + \varepsilon^*_{xx} \tag{3a}$$

$$\varepsilon_{xy} = \frac{1}{2}\left(\frac{\partial U}{\partial y} + \frac{\partial V}{\partial x}\right) + \frac{1}{2}\left(\frac{\partial V}{\partial x}\frac{\partial V}{\partial y} + \frac{\partial W}{\partial x}\frac{\partial W}{\partial y}\right) = \varepsilon^l_{xy} + \varepsilon^*_{xy} \tag{3b}$$

$$\varepsilon_{xz} = \frac{1}{2}\left(\frac{\partial U}{\partial z} + \frac{\partial W}{\partial x}\right) + \frac{1}{2}\left(\frac{\partial V}{\partial x}\frac{\partial V}{\partial z} + \frac{\partial W}{\partial x}\frac{\partial W}{\partial z}\right) = \varepsilon^l_{xz} + \varepsilon^*_{xz} \tag{3c}$$

By using Equations (1)–(3) and considering a tapering geometry, the non-zero linear and nonlinear parts of the strain displacement field are defined as

$$\varepsilon^l_{xx} = u' - yv'' - zw'' - \omega\theta'' \tag{4a}$$

$$\gamma^l_{xz} = 2\varepsilon^l_{xz} = \left(y - \frac{\partial \omega}{\partial z}\right)\theta' \tag{4b}$$

$$\gamma^l_{xy} = 2\varepsilon^l_{xy} = -\left(z + \frac{\partial \omega}{\partial y}\right)\theta' \tag{4c}$$

$$\varepsilon^*_{xx} = \frac{1}{2}\left[v'^2 + w'^2 + r^2\theta'^2\right] + yw'\theta' - zv'\theta' \tag{4d}$$

$$\gamma^*_{xz} = -(v' + \theta'z)\theta \tag{4e}$$

$$\gamma^*_{xy} = (w' + \theta'y)\theta \tag{4f}$$

where $r^2 = y^2 + z^2$. In this study, we consider a compressive axial load P acting at the end of the beam along the z-direction, together with an external bending moment acting around the major principal axis, M^*_y, while assuming a null bending moment M^*_z with respect to the z-axis. The most common cases of normal and shear stress associated with the external bending moment M^*_y and shear force V_z are considered as

$$\sigma^0_{xx} = \frac{P}{A} - \frac{M^*_y}{I_y}z \tag{5a}$$

$$\tau_{xz}^0 = \frac{V_z}{A} = -\frac{M_y^{*\prime}}{A} \tag{5b}$$

where τ_{xz}^0 is the mean value of the shear stress, σ_{xx}^0 stands for the initial normal stress in the cross-section, and A and I_y are the cross-sectional area and second moment of inertia around the y-axis, defined as follows:

$$A = \int_A dA \tag{6a}$$

$$I_y = \int_A z^2 dA \tag{6b}$$

2.2. Constitutive Relations

According to the Eringen nonlocal elasticity model [4], the stress at a point inside a body depends not only on the strain state at that point, but also on the strain states at all other points throughout the body. For homogenous and isotropic elastic solids, the nonlocal stress tensor σ at point x can be defined as

$$\sigma_{ij}(x) = \int_V \alpha(|x' - x|, \tau) C_{ijkl} \varepsilon_{kl}(x') dV(x') \tag{7}$$

where ε_{kl} and C_{ijkl} denote the linear strain components and the elastic stiffness coefficients, respectively. In addition, $\alpha(|x\prime - x|, \tau)$ is the nonlocal kernel function, $|x\prime - x|$ is the Euclidean distance, $\tau = e_0 a / l$ stands for the material parameter, where a is an internal characteristic length (e.g., lattice parameter, C–C bond length or granular distance) and l is an external characteristic length in the nanostructures (e.g., crack length, wavelength), and e_0 is a material constant, which is determined experimentally or in an approximate form by matching the dispersion curves of plane waves with those based on atomic lattice dynamics.

It is possible to express the integral constitutive equation presented in Equation (7) in the following differential constitutive equation:

$$\sigma_{ij} - \mu \nabla^2 \sigma_{ij} = C_{ijkl} \varepsilon_{kl} \tag{8}$$

where ∇^2 is the Laplacian operator and $\mu = (e_0 a)^2$ stands for the nonlocal parameter. For a nonlocal AFG I-beam, the nonlocal constitutive relations can be written as

$$\sigma_{xx} - \mu \frac{\partial^2 \sigma_{xx}}{\partial x^2} = E \varepsilon_{xx}^l \tag{9a}$$

$$\tau_{xy} - \mu \frac{\partial^2 \tau_{xy}}{\partial x^2} = G \gamma_{xy}^l \tag{9b}$$

$$\tau_{xz} - \mu \frac{\partial^2 \tau_{xz}}{\partial x^2} = G \gamma_{xz}^l \tag{9c}$$

where E and G are the elastic and shear moduli, respectively, and σ_{xx}, τ_{xy}, and τ_{xz} denote the Piola–Kirchhoff stress tensor components.

2.3. Equilibrium Equations

The principle of minimum total potential energy is applied to obtain the equilibrium equations together with the boundary conditions. For thin-walled beams, the total potential energy Π is expressed in its variational form by means of the elastic strain energy U_l and the strain energy due to initial stress U_0,

$$\delta \Pi = \delta(U_l + U_0) = 0 \tag{10}$$

Note that in a linear stability context, in the absence of an external force, the external work associated with the applied loads W_e is equal to zero. At the same time, the variational form of the strain energy δU_l is defined as

$$\delta U_l = \int_0^L \int_A \left(\sigma_{xx} \delta \varepsilon_{xx}^l + \tau_{xy} \delta \gamma_{xy}^l + \tau_{xz} \delta \gamma_{xy}^l \right) dA \, dx \tag{11}$$

where L and A stand for the element length and cross-sectional area, respectively, and $\delta \varepsilon_{xx}^l$, $\delta \gamma_{xz}^l$ and $\delta \gamma_{xy}^l$ are the linear parts of the strain tensor in a variational form. By substituting Equation (4a–c) into Equation (11), the virtual elastic strain energy becomes

$$\delta U_l = \int_0^L \int_A \sigma_{xx}(\delta u_0' - y\delta v'' - z\delta w'' - \omega\delta\theta'')dA\,dx + \int_0^L \int_A \tau_{xy}\left(-(z + \tfrac{\partial\omega}{\partial y})\delta\theta'\right)dA\,dx + \int_0^L \int_A \tau_{xz}\left((y - \tfrac{\partial\omega}{\partial z})\delta\theta'\right)dA\,dx \tag{12}$$

By integration over the cross-sectional area, we get

$$\delta U_l = \int_L \left(N\delta u_0' + M_z\delta v'' - M_y\delta w'' + B_\omega\delta\theta'' \right)dx + \int_0^L \left(M_{sv}\delta\theta' \right)dx \tag{13}$$

where N is the axial force, M_y and M_z denote the two bending moments, B_ω is the bi-moment, and M_{sv} is the St. Venant torsional moment. These stress resultants in Equation (13) are defined as

$$N = \int_A \sigma_{xx} dA \tag{14a}$$

$$M_y = \int_A \sigma_{xx} z \, dA \tag{14b}$$

$$M_z = -\int_A \sigma_{xx} y \, dA \tag{14c}$$

$$B_\omega = -\int_A \sigma_{xx} \omega \, dA \tag{14d}$$

$$M_{sv} = \int_A \left(\tau_{xz}(y - \frac{\partial\omega}{\partial z}) - \tau_{xy}(z + \frac{\partial\omega}{\partial y}) \right) dA \tag{14e}$$

Moreover, the variation in the strain energy due to the initial stresses can be stated as

$$\delta U_0 = \int_0^L \int_A \left(\sigma_{xx}^0 \delta\varepsilon_{xx}^* + \tau_{xy}^0 \delta\gamma_{xy}^* + \tau_{xz}^0 \delta\gamma_{xz}^* \right) dA \, dx \tag{15}$$

By introducing the first variation in the nonlinear strain-displacement relations, defined by Equation (4d–f), and the initial stresses (5a,b) in Equation (15), we get the following relation:

$$\delta U_0 = \int_0^L \int_A \left(\frac{P}{A} - \frac{M_y^*}{I_y}z\right)\left(v'\delta v' + w'\delta w' + r^2\theta'\delta\theta' + y\theta'\delta w\prime + yw\prime\delta\theta\prime - z\theta\prime\delta v\prime - zv\prime\delta\theta\prime\right)dA\,dx$$
$$+ \int_0^L \int_A \left(-\frac{M_y^{*\prime}}{A}\right)\left(-\theta\delta v' - v'\delta\theta - z\theta\delta\theta' - z\theta'\delta\theta\right)dA\,dx \tag{16}$$

At this stage, by integrating Equation (16) over the cross-section, the variation in the strain energy due to the initial stresses takes the following final form:

$$\delta U_0 = \int_0^L \left(P(v'\delta v' + w'\delta w' + \frac{I_y + I_z}{A}\theta'\delta\theta') \right)dx + \int_0^L \left(M_y^*(\theta\prime\delta v\prime + v\prime\delta\theta\prime) \right)dx + \int_0^L \left(M_y^{*\prime}(\theta\delta v' + v'\delta\theta) \right)dx \tag{17}$$

or equivalently

$$\delta U_0 = \int_0^L \left(Pv'\delta v' + Pw'\delta w' + r_0^2\theta'\delta\theta' \right)dx + \int_0^L \left(-M_y^*v''\delta\theta - M_y^*\theta\delta v'' \right)dx \tag{18}$$

In Equation (18), I_z is the second moment of inertia around the z-axis and r_0 is the polar radius gyration around the centroid, given by

$$I_z = \int_A y^2 dA, \ r_o = \sqrt{\frac{I_y + I_z}{A}}$$

By introducing Equations (13) and (18) into Equation (10) and setting the coefficients of $\delta u_0, \delta v, \delta w, \delta \theta$, as to zero, we obtain the equilibrium equations

$$N' = 0 \tag{19a}$$

$$-M_y'' - (Pw')' = 0 \tag{19b}$$

$$M_z'' - (M_y^*\theta)'' - (Pv')' = 0 \tag{19c}$$

$$M_\omega'' - M_y^* v'' - M_{sv}' - (Pr_O^2 \theta')' = 0 \tag{19d}$$

under the following boundary conditions

$$
\begin{aligned}
N = 0 &\quad \text{or} \quad \delta u_0 = 0 \\
-M_y = 0 &\quad \text{or} \quad \delta w' = 0 \\
M_y' + Pw' = 0 &\quad \text{or} \quad \delta w = 0 \\
M_z - M_y^*\theta = 0 &\quad \text{or} \quad \delta v\prime = 0 \\
-M_z' + (M_y^*\theta)' + Pv' = 0 &\quad \text{or} \quad \delta v = 0 \\
-B_\omega = 0 &\quad \text{or} \quad \delta\theta' = 0 \\
B_\omega' + M_{sv} + Pr_O^2\theta' = 0 &\quad \text{or} \quad \delta\theta = 0
\end{aligned}
\tag{20}
$$

By substituting Equation (4a–c) into Equation (9) and the subsequent results into Equation (14), the stress resultants are obtained as

$$N - \mu\frac{\partial^2 N}{\partial x^2} = EAu_0' \tag{21a}$$

$$M_y - \mu\frac{\partial^2 M_y}{\partial x^2} = -EI_y w'' \tag{21b}$$

$$M_z - \mu\frac{\partial^2 M_z}{\partial x^2} = EI_z v'' \tag{21c}$$

$$B_\omega - \mu\frac{\partial^2 B_\omega}{\partial x^2} = EI_\omega \theta'' \tag{21d}$$

$$M_{sv} - \mu\frac{\partial^2 M_{sv}}{\partial x^2} = GJ\theta' \tag{21e}$$

In the previous expressions, J and I_ω are the St. Venant torsion and warping constants, defined as

$$I_\omega = \int_A \omega^2 dA, \tag{22a}$$

$$J = \int_A \left((y - \frac{\partial\omega}{\partial z})^2 + (z + \frac{\partial\omega}{\partial y})^2 \right) dA \tag{22b}$$

This study is established in the context of small displacements and deformations. According to the linear stability, the nonlinear terms are also disregarded in the equilibrium equations. Based on these assumptions, the system of equilibrium equations for tapered I-beams under a nonlocal theory are finally derived by placing Equation (21) into Equation (19)

$$\left(EAu'_0\right)' = 0 \tag{23a}$$

$$\left(EI_y w''\right)'' + \mu(Pw')''' - (Pw')' = 0 \tag{23b}$$

$$\left(EI_z v''\right)'' + \mu(M_y^*\theta)''''- (M_y^*\theta)'' + \mu(Pv')''' - (Pv')' = 0 \tag{23c}$$

$$\left(EI_\omega \theta''\right)'' - (GJ\theta')' + \mu(M_y^*v'')'' - M_y^*v'' + \mu(Pr_O^2\theta')''' - (Pr_O^2\theta')' = 0 \tag{23d}$$

The related boundary conditions at the ends of the thin-walled nanobeam can be expressed as

$$\left(EAu'_0\right)' = 0 \text{ or } \delta u_0 = 0 \tag{24}$$

$$EI_y w'' = 0 \text{ or } \delta w' = 0$$

$$-(EI_y w'')' - \mu(Pw')'' + Pw' = 0 \text{ or } \delta w = 0$$

$$EI_z v'' - M_y^*\theta + \mu(M_y^*\theta)'' = 0 \text{ or } \delta v' = 0$$

$$-(EI_z v'')' - \mu(M_y^*\theta)''' - \mu(Pv')'' + (M_y^*\theta)' + Pv' = 0 \text{ or } \delta v = 0$$

$$EI_\omega \theta'' = 0 \text{ or } \delta \theta' = 0$$

$$-(EI_\omega \theta'')' + GJ\theta' - \mu(M_y^*v'')' + M_y^*v' - \mu(Pr_O^2\theta')'' + (Pr_O^2\theta') = 0 \text{ or } \delta \theta = 0$$

In the following section, a numerical solution procedure based on the DQM is applied to solve the governing equations for the flexural–torsional buckling of AFG nanobeams with varying I-sections, as has been successfully carried out in the literature for a large variety of problems [96–103].

3. Numerical Solution Method

Due to the varying cross-sectional mechanical properties, the resulting flexural–torsional stability Equation (23a–d) for I-tapered nanobeams represent a system of three-coupled fourth-order differential equations with variable coefficients. Under these conditions, it is not possible to accurately estimate a general and straightforward closed-form solution. For such complicated problems, the DQM-based approach, as proposed for the first time by Bellman and Casti [96], is here employed as an efficient and easy tool to solve the coupled differential equations of the problem in a strong form. The basic concept of the proposed method relies on the possibility of discretizing the derivatives of a function with respect to a variable in differential equations at some fixed collocation points by means of a weighted linear summation of the function's values at its adjacent points. The governing equations, together with the associated boundary conditions, are thus transformed into a set of linear algebraic equations, which can be solved with the aid of a computational algorithm to derive an approximate solution for continuous differential equations. To this end, it is necessary to divide the computational region into a fixed number of grid points spanning the solution domain. The accuracy of this numerical approach depends on the number and types of selected sampling points, as also discussed in Refs. [97–103]. One of the best options for the sampling points in the stability and vibration analysis is the Chebyshev–Gauss–Lobatto points:

$$x_i = \frac{L}{2}\left[1 - \cos\left(\frac{i-1}{N-1}\pi\right)\right], \quad \text{if } 0 \le x \le L \quad i = 1, 2, \ldots, N \tag{25}$$

where N is the total number of grid points in the longitudinal direction. According to DQM, the mth-order derivative of a function $f(\zeta)$ at a fixed grid point ζ_i can be approximated as

$$\left.\frac{d^m f}{d\zeta^m}\right|_{\zeta=\zeta_i} = \sum_{j=1}^{N} A_{ij}^{(m)} f(\zeta_j) \quad \text{for } i = 1, 2, \ldots, N \tag{26}$$

where $f(\xi_j)$ refers to the functional value at grid points ξ_j $(i = 1, 2, \ldots, N)$, and $A_{ij}^{(m)}$ is the weighting coefficient for the mth-order derivative. The first-order derivative of the weighting coefficient $A_{ij}^{(1)}$ is computed by the following algebraic formulation based on the Lagrangian interpolation polynomials,

$$A_{ij}^{(1)} = \begin{cases} \dfrac{M(\xi_i)}{(\xi_i - \xi_j)M(\xi_j)} & for \ i \neq j \\ -\sum\limits_{k=1, k \neq i}^{N} A_{ik}^{(1)} & for \ i = j \end{cases} \quad i, j = 1, 2, \ldots, N \tag{27}$$

where

$$M(\xi_i) = \prod_{j=1, j \neq i}^{N} (\xi_i - \xi_j) \ for \ i = 1, 2, \ldots, N \tag{28}$$

The higher-order DQM weighting coefficients can be acquired from the first-order ones, as follows:

$$A_{ij}^{(m)} = A_{ij}^{(1)} A_{ij}^{(m-1)} \quad 2 \leq m \leq N - 1 \tag{29}$$

In order to solve the stability equation by means of the differential quadrature approach, a dimensionless variable ($\xi = x/L$) is introduced. By the expansion of Equation (23), the governing equations of the problem take the following final discrete form:

$$E(\xi_j)I_y(\xi_j)(\sum_{j=1}^{N} A_{ij}^{(4)} w_j) + 2(E(\xi_j)I_y'(\xi_j) + E'(\xi_j)I_y(\xi_j))(\sum_{j=1}^{N} A_{ij}^{(3)} w_j)$$

$$+ (E''(\xi_j)I_y(\xi_j) + 2E'(\xi_j)I_y'(\xi_j) + E(\xi_j)I_y''(\xi_j))(\sum_{j=1}^{N} A_{ij}^{(2)} w_j) + \mu P(\sum_{j=1}^{N} A_{ij}^{(4)} w_j) - L^2 P(\sum_{j=1}^{N} A_{ij}^{(2)} w_j) = 0 \tag{30a}$$

$$E(\xi_j)I_z(\xi_j)(\sum_{j=1}^{N} A_{ij}^{(4)} v_j) + 2(E(\xi_j)I_z'(\xi_j) + E'(\xi_j)I_z(\xi_j))(\sum_{j=1}^{N} A_{ij}^{(3)} v_j)$$

$$+ (E''(\xi_j)I_z(\xi_j) + 2E'(\xi_j)I_z'(\xi_j) + E(\xi_j)I_z''(\xi_j))(\sum_{j=1}^{N} A_{ij}^{(2)} v_j) + \mu P(\sum_{j=1}^{N} A_{ij}^{(4)} v_j) - L^2 P(\sum_{j=1}^{N} A_{ij}^{(2)} v_j)$$

$$+ \mu M_y^*(\xi_j)(\sum_{j=1}^{N} A_{ij}^{(4)} \theta_j) + 4\mu M_y^{*\prime}(\xi_j)(\sum_{j=1}^{N} A_{ij}^{(3)} \theta_j) + (6\mu M_y^{*\prime\prime}(\xi_j) - L^2 M_y^*(\xi_j))(\sum_{j=1}^{N} A_{ij}^{(2)} \theta_j)$$

$$+ (4\mu M_y^{*\prime\prime\prime}(\xi_j) - 2L^2 M_y^{*\prime}(\xi_j))(\sum_{j=1}^{N} A_{ij}^{(1)} \theta_j) + (\mu M_y^{*4}(\xi_j) - L^2 M_y^{*\prime\prime}(\xi_j))\theta_j = 0 \tag{30b}$$

$$E(\xi_j)I_\omega(\xi_j)(\sum_{j=1}^{N} A_{ij}^{(4)} \theta_j) + 2(E(\xi_j)I_\omega'(\xi_j) + E'(\xi_j)I_\omega(\xi_j))(\sum_{j=1}^{N} A_{ij}^{(3)} \theta_j)$$

$$+ (E''(\xi_j)I_\omega(\xi_j) + 2E'(\xi_j)I_\omega'(\xi_j) + E(\xi_j)I_\omega''(\xi_j) - L^2 G(\xi_j)J(\xi_j))(\sum_{j=1}^{N} A_{ij}^{(2)} \theta_j)$$

$$- L^2(G'(\xi_j)J(\xi_j) + G(\xi_j)J'(\xi_j))(\sum_{j=1}^{N} A_{ij}^{(1)} \theta_j)$$

$$+ \mu PR_c(\xi_j)(\sum_{j=1}^{N} A_{ij}^{(4)} \theta_j) + 3\mu PR_o'(\xi_j)(\sum_{j=1}^{N} A_{ij}^{(3)} \theta_j) + P(3\mu R_c'(\xi_j) - L^2 R_o(\xi_j))(\sum_{j=1}^{N} A_{ij}^{(2)} \theta_j)$$

$$+ P(\mu R_o''(\xi_j) - L^2 R_o'(\xi_j))(\sum_{j=1}^{N} A_{ij}^{(1)} \theta_j) + \mu M_y^*(\xi_j)(\sum_{j=1}^{N} A_{ij}^{(4)} v_j) + 2\mu M_y^{*\prime}(\xi_j)(\sum_{j=1}^{N} A_{ij}^{(3)} v_j)$$

$$+ (\mu M_y^{*\prime\prime}(\xi_j) - L^2 M_y^*(\xi_j))(\sum_{j=1}^{N} A_{ij}^{(2)} v_j) = 0 \tag{30c}$$

where $R_o = r_o^2$ in Equation (30c).

By rewriting the problem in matrix form, we get the following relation,

$$\left(\begin{bmatrix} [K_{ww}] & [0] & [0] \\ [0] & [K_{vv}] & [0] \\ [0] & [0] & [K_{\theta\theta}] \end{bmatrix}_{3N\times3N} + \begin{bmatrix} [P_{ww}] & [0] & [0] \\ [0] & [P_{vv}] & [0] \\ [0] & [0] & [P_{\theta\theta}] \end{bmatrix}_{3N\times3N} + \begin{bmatrix} [0] & [0] & [0] \\ [0] & [0] & [M_{v\theta}] \\ [0] & [M_{\theta v}] & [0] \end{bmatrix}_{3N\times3N}\right) \times \left\{\begin{matrix} \{w\} \\ \{v\} \\ \{\theta\} \end{matrix}\right\}_{3N\times1} = \left\{\begin{matrix} \{0\} \\ \{0\} \\ \{0\} \end{matrix}\right\}_{3N\times1} \tag{31}$$

where

$$[K_{ww}] = [a^1][A]^{(4)} + [b^1][A]^{(3)} + [c^1][A]^{(2)}$$

$$[P_{ww}] = P(\mu[A]^{(4)} - L^2[A]^{(2)})$$

$$[K_{vv}] = [a^2][A]^{(4)} + [b^2][A]^{(3)} + [c^2][A]^{(2)}$$

$$[P_{vv}] = P(\mu[A]^{(4)} - L^2[A]^{(2)})$$

$$[M_{v\theta}] = [i^2][A]^{(4)} + [j^2][A]^{(3)} + [k^2][A]^{(2)} + [l^2][A]^{(1)} + [m^2]$$

$$[K_{\theta\theta}] = [a^3][A]^{(4)} + [b^3][A]^{(3)} + [c^3][A]^{(2)} - [d^3][A]^{(1)}$$

$$[P_{\theta\theta}] = P([e^3][A]^{(4)} + [f^3][A]^{(3)} + [g^3][A]^{(2)} + [h^3][A]^{(1)})$$

$$[M_{\theta v}] = [i^3][A]^{(4)} + [j^3][A]^{(3)} + [k^3][A]^{(2)} \tag{32}$$

in which

$$a^1_{jk} = (EI_y|_{\xi=\xi_j})\delta_{jk}; \; b^1_{jk} = (2(EI'_y + E'I_y)|_{\xi=\xi_j})\delta_{jk}; \; c^1_{jk} = ((E''I_y + 2E'I'_y + EI''_y)|_{\xi=\xi_j})\delta_{jk} \tag{33}$$

$$a^2_{jk} = (EI_z|_{\xi=\xi_j})\delta_{jk}; \; b^2_{jk} = (2(EI'_z + E'I_z)|_{\xi=\xi_j})\delta_{jk}; \; c^2_{jk} = ((E''I_z + 2E'I'_z + EI''_z)|_{\xi=\xi_j})\delta_{jk}$$

$$i^2_{jk} = (\mu M^*_y|_{\xi=\xi_j})\delta_{jk}; \; j^2_{jk} = (4\mu M^{*'}_y|_{\xi=\xi_j})\delta_{jk}; \; k^2_{jk} = ((6\mu M^{*''}_y - L^2 M^*_y)|_{\xi=\xi_j})\delta_{jk}$$

$$l^2_{jk} = ((4\mu M^{*''}_y - 2L^2 M^{*'}_y)|_{\xi=\xi_j})\delta_{jk}; \; m^2_{jk} = ((\mu M^{*4}_y - L^2 M^{*''}_y)|_{\xi=\xi_j})\delta_{jk}$$

$$a^3_{jk} = (EI_\omega|_{\xi=\xi_j})\delta_{jk}; \; b^3_{jk} = (2(EI'_\omega + E'I_\omega)|_{\xi=\xi_j})\delta_{jk}; \; c^3_{jk} = ((E''I_\omega + 2E'I'_\omega + EI''_\omega - L^2 GJ)|_{\xi=\xi_j})\delta_{jk};$$

$$d^3_{jk} = (L^2(G'J + GJ')|_{\xi=\xi_j})\delta_{jk}; \; e^3_{jk} = P(\mu R_0|_{\xi=\xi_j})\delta_{jk}; \; f^3_{jk} = P(3\mu R'_0|_{\xi=\xi_j})\delta_{jk};$$

$$g^3_{jk} = P((3\mu R'_0 - L^2 R_0)|_{\xi=\xi_j})\delta_{jk}; \; h^3_{jk} = P((\mu R''_0 - L^2 R'''_0)|_{\xi=\xi_j})\delta_{jk}$$

$$i^3_{jk} = (\mu M^*_y|_{\xi=\xi_j})\delta_{jk}; \; j^3_{jk} = (2\mu M^{*'}_y|_{\xi=\xi_j})\delta_{jk}; \; k^3_{jk} = ((\mu M^{*''}_y - L^2 M^*_y)|_{\xi=\xi_j})\delta_{jk}$$

and δ_{jk} is the Kronecker delta, defined as

$$\delta_{jk} = \begin{cases} 0 & if \; j \neq k; \\ 1 & if \; j = k. \end{cases} \tag{34}$$

In Equation (31), the displacement vectors and the torsion angle vector are defined as

$$\{w\}_{N\times1} = \left\{\begin{matrix} w_1 & w_2 & \dots & w_N \end{matrix}\right\}^T; \; \{v\}_{N\times1} = \left\{\begin{matrix} v_1 & v_2 & \dots & v_N \end{matrix}\right\}^T;$$
$$\{\theta\}_{N\times1} = \left\{\begin{matrix} \theta_1 & \theta_2 & \dots & \theta_N \end{matrix}\right\}^T \tag{35}$$

The simple form of the final equation, Equation (31), can be stated as

$$([K] - \lambda([P] + [M]))_{3N\times3N}\{d\}_{3N\times1} = \{0\}_{3N\times1} \tag{36}$$

or

$$([K] - \lambda[K_G])\{d\} = \{0\} \tag{37}$$

in which

$$[K_G] = [P] + [M] \tag{38a}$$

$$\{d\} = \left\{ \begin{array}{c} \{w\} \\ \{v\} \\ \{\theta\} \end{array} \right\} \tag{38b}$$

$[K]$ and $[K_G]$ are $3N \times 3N$ matrices, λ is the eigenvalues and $\{d\}$ is the related eigenvectors. After the implementation of the boundary conditions, we compute the flexural–torsional buckling load from Equation (37), together with the associated vertical and lateral deflections and the twist angles of the AFG nanobeams.

4. Numerical Examples

In this section, we perform a parametric investigation to assess the sensitivity of the linear stability of AFG thin-walled nanobeam-columns (with a variable I-section and simply supported boundary conditions) to different material properties, as well as to different web and flange tapering parameters, mode numbers, nonlocal parameters, and axial load eccentricities. In what follows, we use the subscripts $(\bullet)_0$ and $(\bullet)_1$ to define the mechanical and geometrical properties of beams in their left ($x = 0$, $\xi = 0$) and right ($x = L$, $\xi = 1$) supports, respectively. The dimensionless buckling load parameter is determined as

$$P_{nor} = \frac{P_{cr}L^2}{E_0 I_{Z0}} \tag{39}$$

which accounts for simply supported, tapered beams with I-sections subjected to a compressive axial force. In this regard, it is presumed that the widths of both flanges, b_0, and the web height, d_0, of the I-section on the left side increase linearly up to $b_1 = (1 + \beta)b_0$ and $d_1 = (1 + \alpha)d_0$ on the right side (Figure 2). Thus, the flanges and web tapering ratios are defined as $\beta = b_1/b_0 - 1$ and $\alpha = d_1/d_0 - 1$, respectively. Note that these two parameters (α, β) are non-negative variables and can change simultaneously or separately. At the same time, by equating (α, β) to zero, we revert to I-beams with a uniform cross-section. The geometrical schemes and dimensionless parameters are depicted in Figure 2. To perform the flexural–torsional buckling analysis, it is supposed that the compressive axial load is applied at three different positions: the top flange (TF) of the left side (i.e., for $x = 0$), the centroid, and the TF of the right side (i.e., for $x = L$).

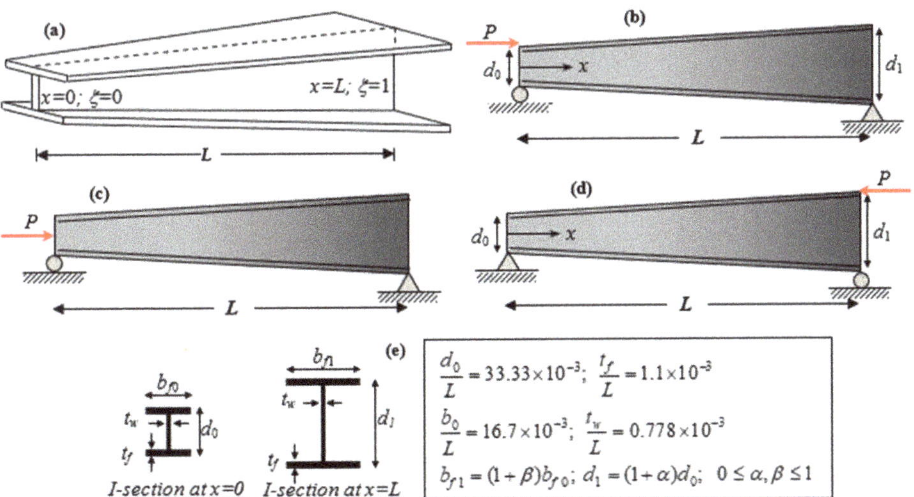

Figure 2. (**a**) Schematic representation of double-tapered beam with I-sections. (**b**) Axial load on the TF at $x = 0$ (**c**) Axial load on the centroid. (**d**) Axial load on the TF at $x = L$ (**e**) Geometrical properties.

For the same benchmark, the beams feature axially varying materials, ranging between pure ceramic on the side end and pure metal on the right side, according to a simple power-law function. More specifically, the ceramic phase is made of alumina (Al_2O_3) with an equivalent Young's modulus $E_0 = 380$ GPa, whereas the metal phase is aluminum (Al) with an equivalent Young's modulus $E_1 = 70$ GPa, without considering the exact grain sizes and shapes of each material constituent. This means that the modulus of elasticity at an arbitrary coordinate is defined as

$$E(\xi) = E_0 + (E_1 - E_0)\xi^m \tag{40}$$

where the power-law index m assumes a positive value, and is zero only in a pure metal member.

We carry out a preliminary study aimed at defining the appropriate number of grid points within the domain to yield accurate results in terms of flexural–torsional buckling load. In the absence of further numerical nonlocal studies on the same thin-walled examples, the accuracy of our formulation is checked by comparing our results with predictions based on a classical finite element method, as performed via the commercial ANSYS code [104]. In detail, we evaluate the lowest values of the dimensionless buckling parameter (P_{nor}) for the same structure made of pure alumina with three different loading positions and different tapering ratios, $\alpha = \beta = 0 \div 1$ by steps of 0.2 versus an increased number of grid points N. The main results are summarized in Table 1, where it seems that a total number of grid points $N = 20$ is sufficient to obtain the lowest normalized buckling load for different axial load positions and non-uniformity parameters. Based on results in Table 1, we can observe the good agreement between our mathematical DQM-based formulation and predictions made via the ANSYS code [104] for each selected loading case.

Table 1. Dimensionless buckling load (P_{nor}) for local tapered homogenous I-beams (alumina) with different tapering parameters and loading positions.

Axial Load Position	$\alpha = \beta$	DQM					ANSYS [104]
		Number of Points along x-Direction					
		$n = 5$	$n = 10$	$n = 15$	$n = 20$	$n = 30$	
Centroid	0	9.824	9.870	9.870	9.870	9.870	9.866
	0.2	12.970	13.006	13.006	13.006	13.006	12.997
	0.4	16.550	16.494	16.494	16.494	16.494	16.466
	0.6	20.647	20.326	20.326	20.326	20.326	20.276
	0.8	25.426	24.497	24.497	24.497	24.497	24.414
	1.0	31.184	29.004	29.003	29.003	29.003	27.605
TF of left end section	0	9.213	9.248	9.248	9.248	9.248	9.274
	0.2	11.974	11.964	11.964	11.964	11.964	12.000
	0.4	15.050	14.895	14.895	14.895	14.895	14.907
	0.6	18.570	18.029	18.029	18.029	18.029	18.050
	0.8	22.915	21.358	21.357	21.357	21.357	21.399
	1.0	29.257	24.877	24.876	24.876	24.876	24.956
TF of right end section	0	9.213	9.248	9.248	9.248	9.248	9.274
	0.2	11.805	11.862	11.862	11.862	11.862	11.922
	0.4	14.557	14.610	14.610	14.610	14.610	14.690
	0.6	17.524	17.469	17.469	17.469	17.469	17.638
	0.8	20.912	20.429	20.429	20.429	20.429	20.737
	1.0	25.283	23.488	23.489	23.489	23.489	23.992

After the validation phase of the model, we continue with a systematic study of the flexural–torsional buckling of AFG nanobeams with different input parameters, such as eccentric axial load, web and flange non-uniformity parameters, gradient index, mode number and nonlocality parameter. In order to assess the linear stability strength of AFG nanobeams with varying I-sections, we compute the lowest normalized flexural–torsional buckling loads (P_{nor}) of AFG tapered thin-walled nanobeams subjected to simply supported end conditions, as reported in Table 2, for different tapering ratios ($\alpha = \beta = 0$, 0.3, 0.6, 0.9), material compositions (power-law exponent), nonlocal parameters ($\mu = 0$ and 2), and three different loading positions. The compressive axial load can be applied on the TF of the left side ($x = 0$), at the centroid, and on the TF of the right side ($x = L$). In Figures 3–5, we represent the variation in P_{nor} depending on Eringen's nonlocal parameters (ranging from 0 to 3) for thin-walled beams with homogenous materials or an FG beam with different gradient indexes $m = 0.6$, 1.3 and 2, while varying the tapering ratios from 0 to 0.9 and assuming three axial load positions, namely, on the TF for $x = 0$, on the centroid, and the TF for $x = L$. In this case, we consider a non-uniform beam with equal web height and flange width tapering ratios, $\alpha = \beta$.

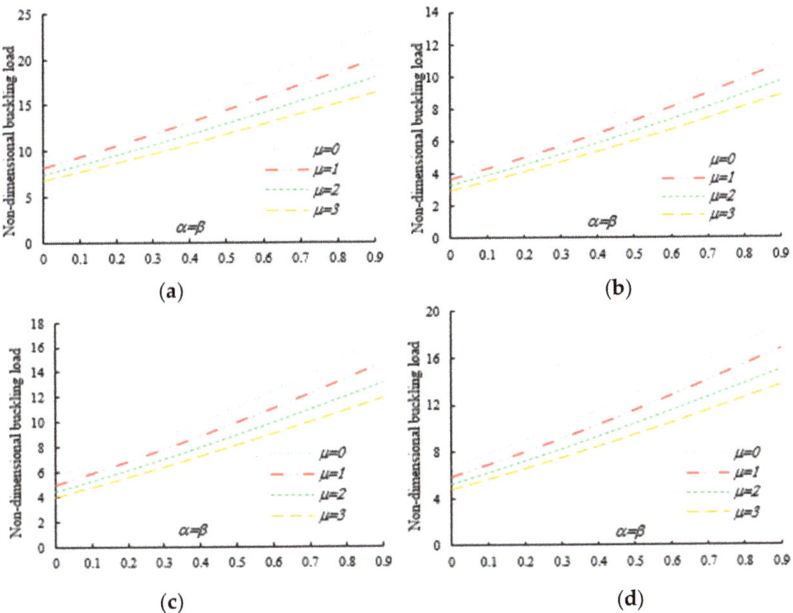

Figure 3. Variation in the flexural–torsional buckling load (P_{nor}) of I-tapered nanobeams with varying tapering and nonlocality parameters—Figure 2. Effect of the axial load eccentricity, material graduation and tapering parameter on the normalized buckling load P_{nor} for simply supported thin-walled nanobeams subjected to a constant compressive load with two different nonlocal parameters: (**a**)Homogeneous (Alumina); (**b**) $m = 0.6$; (**c**) $m = 1.3$; (**d**) $m = 2$.

In Tables 3–5, we list the magnitude of the normalized flexural–torsional buckling parameter, P_{nor}, for various combinations of web height and flange width tapering ratio, β and α, and nonlocal parameters ($\mu = 0$, 1 and 3) with different non-homogenous indices ($m = 0.6$, 1.2 and 1.8). The contribution of a possible axial load eccentricity at the cross-section centroid on the buckling resistance is also taken into account. The normalized buckling parameters are respectively illustrated in Tables 3 and 4 for load positions on the TF at $x = 0$ and on the shear center, as well as in Table 5 for a load position on the TF at $x = L$.

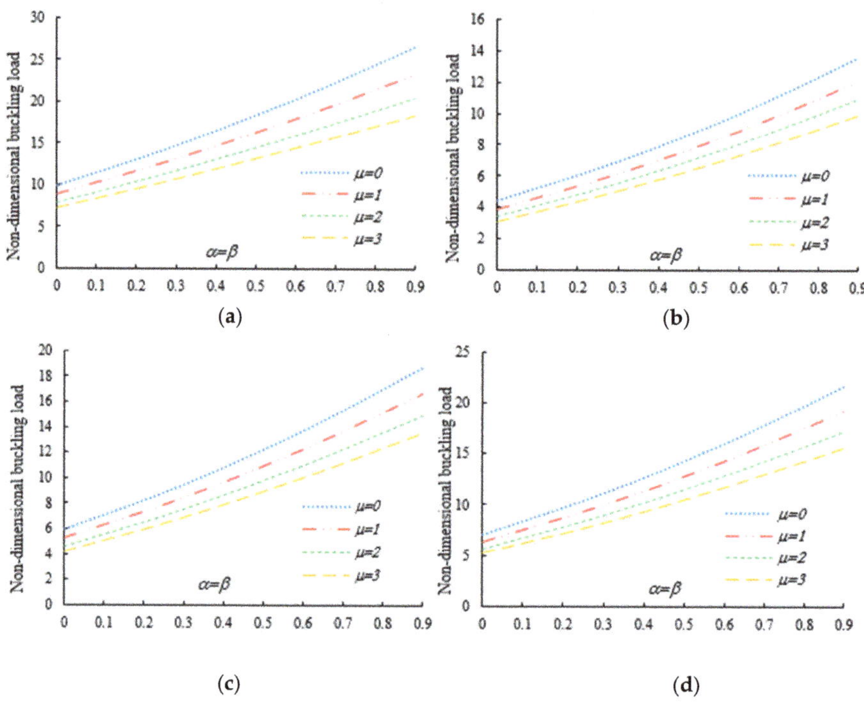

Figure 4. Variation in the flexural buckling load P_{nor} of I-tapered nanobeams with variations in the tapering and nonlocality parameters for different material indexes (axial load on the centroid) (**a**) Homogeneous (Alumina); (**b**) $m = 0.6$; (**c**) $m = 1.3$; (**d**) $m = 2$.

Table 2. Effect of the axial load eccentricity, material graduation and tapering parameter on the normalized buckling load (P_{nor}) for simply supported thin-walled nanobeams subjected to a constant compressive load with two different nonlocal parameters.

μ (nm²)	$\alpha = \beta$	Axial Load on the TF at $x = 0$				Axial Load on the Centroid				Axial Load on the TF at $x = L$			
		Homogeneous	$m = 0.8$	$m = 1.6$	$m = 2.4$	Homogeneous	$m = 0.8$	$m = 1.6$	$m = 2.4$	Homogeneous	$m = 0.8$	$m = 1.6$	$m = 2.4$
0	0.0	9.248	4.602	6.165	7.151	9.870	4.816	6.489	7.571	9.248	4.383	5.863	6.843
	0.3	13.416	7.140	9.516	10.879	14.711	7.641	10.272	11.833	13.232	6.596	8.804	10.165
	0.6	18.070	10.143	13.401	15.154	20.338	11.104	14.836	16.929	17.506	9.086	12.071	13.839
	0.9	23.186	13.578	17.787	19.927	26.738	15.194	20.169	22.823	22.029	11.827	15.648	17.827
2.0	0.0	7.453	3.595	4.836	5.684	7.936	3.756	5.088	6.028	7.453	3.389	4.714	5.705
	0.3	10.646	5.744	7.658	8.754	11.750	6.102	8.204	9.471	10.679	5.180	6.899	8.010
	0.6	14.010	8.171	10.752	12.080	15.996	8.922	11.902	13.547	14.110	7.261	9.655	11.113
	0.9	17.488	10.850	14.071	15.572	20.591	12.178	16.080	18.061	17.713	9.558	12.662	14.450

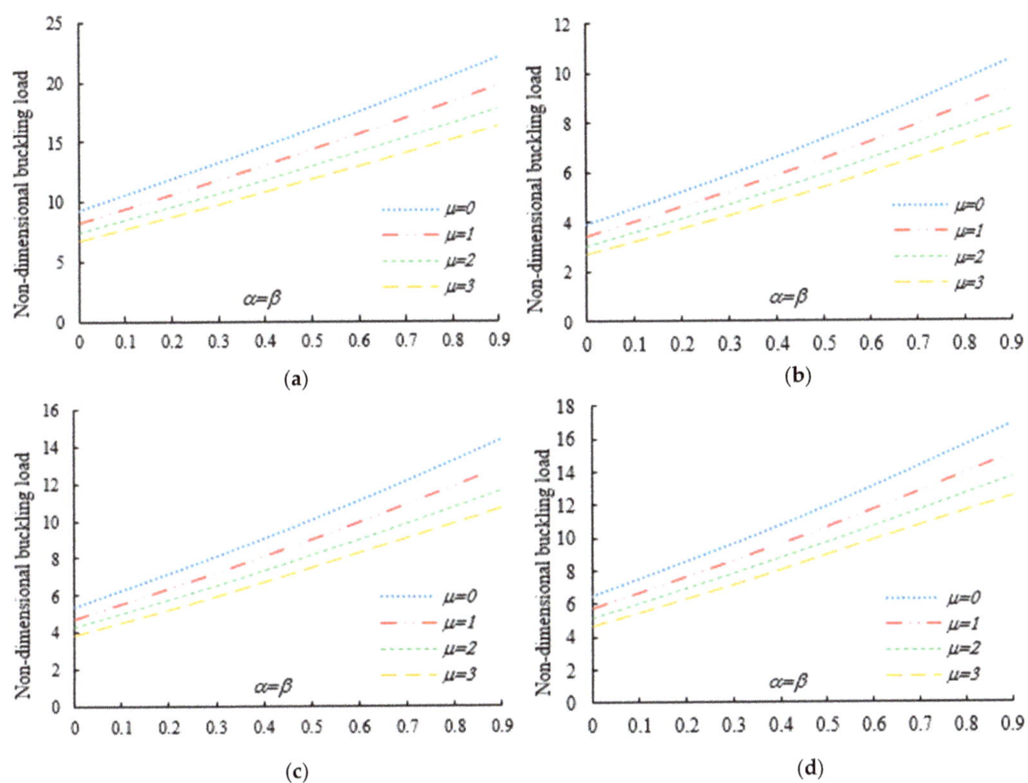

Figure 5. Variation in the flexural–torsional buckling load P_{nor} of I-tapered nanobeams with variations in the tapering and nonlocality parameters for different material indexes (axial load on the TF at $x = L$): (**a**) Homogeneous (Alumina); (**b**) $m = 0.6$; (**c**) $m = 1.3$; (**d**) $m = 2$.

Table 3. Effect of the power-law exponent and tapering parameter on the normalized flexural–torsional buckling load (P_{nor}) of simply supported thin-walled nanobeams with different nonlocal parameters (axial load applied on the TF at $x = 0$).

μ (nm^2)	α	$m = 0.6$				$m = 1.2$				$m = 1.8$			
		$\beta = 0$	$\beta = 0.2$	$\beta = 0.5$	$\beta = 0.8$	$\beta = 0$	$\beta = 0.2$	$\beta = 0.5$	$\beta = 0.8$	$\beta = 0$	$\beta = 0.2$	$\beta = 0.5$	$\beta = 0.8$
0	0.0	4.057	5.488	7.943	10.725	5.483	7.413	10.693	14.373	6.462	8.685	12.418	16.559
	0.2	4.065	5.503	7.973	10.778	5.496	7.435	10.739	14.454	6.479	8.713	12.475	16.658
	0.5	4.077	5.524	8.015	10.852	5.514	7.467	10.802	14.565	6.502	8.754	12.554	16.793
	0.8	4.089	5.543	8.053	10.919	5.531	7.496	10.859	14.664	6.524	8.790	12.625	16.914
1.0	0.0	3.588	4.886	7.089	9.543	4.842	6.598	9.539	12.765	5.715	7.735	11.062	14.650
	0.2	3.595	4.899	7.115	9.590	4.852	6.617	9.578	12.835	5.729	7.760	11.111	14.736
	0.5	3.605	4.916	7.151	9.654	4.868	6.643	9.632	12.932	5.748	7.793	11.179	14.854
	0.8	3.614	4.933	7.184	9.713	4.881	6.668	9.682	13.020	5.765	7.824	11.242	14.962
3.0	0.0	2.899	4.008	5.853	7.848	3.886	5.405	7.871	10.461	4.586	6.346	9.109	11.916
	0.2	2.904	4.017	5.873	7.885	3.893	5.418	7.901	10.516	4.594	6.363	9.147	11.985
	0.5	2.911	4.030	5.901	7.936	3.902	5.437	7.943	10.594	4.605	6.387	9.200	12.083
	0.8	2.917	4.042	5.927	7.985	3.910	5.454	7.982	10.667	4.615	6.409	9.250	12.173

Table 4. Effect of the power-law exponent and tapering parameter on the normalized flexural–torsional buckling load (P_{nor}) of simply supported thin-walled nanobeams with different nonlocal parameters (axial load applied on the centroid).

μ (nm^2)	α	$m = 0.6$				$m = 1.2$				$m = 1.8$			
		$\beta = 0$	$\beta = 0.2$	$\beta = 0.5$	$\beta = 0.8$	$\beta = 0$	$\beta = 0.2$	$\beta = 0.5$	$\beta = 0.8$	$\beta = 0$	$\beta = 0.2$	$\beta = 0.5$	$\beta = 0.8$
0	0.0	4.243	5.837	8.689	12.096	5.756	7.928	11.796	16.392	6.815	9.345	13.814	19.077
	0.2	4.244	5.838	8.690	12.097	5.757	7.929	11.797	16.392	6.816	9.346	13.815	19.078
	0.5	4.245	5.839	8.691	12.098	5.759	7.930	11.798	16.394	6.817	9.348	13.817	19.079
	0.8	4.246	5.840	8.692	12.099	5.760	7.931	11.800	16.395	6.819	9.349	13.818	19.081
1.0	0.0	3.736	5.177	7.741	10.775	5.054	7.023	10.504	14.589	5.989	8.287	12.299	16.949
	0.2	3.736	5.178	7.742	10.776	5.055	7.024	10.505	14.589	5.990	8.288	12.300	16.950
	0.5	3.737	5.179	7.743	10.777	5.056	7.026	10.506	14.590	5.992	8.289	12.302	16.951
	0.8	3.738	5.180	7.744	10.777	5.058	7.027	10.507	14.592	5.993	8.291	12.303	16.953
3.0	0.0	2.990	4.210	6.353	8.840	4.000	5.688	8.610	11.945	4.725	6.716	10.077	13.823
	0.2	2.990	4.210	6.353	8.840	4.001	5.689	8.611	11.946	4.726	6.717	10.078	13.823
	0.5	2.991	4.211	6.354	8.841	4.002	5.690	8.612	11.947	4.727	6.718	10.079	13.824
	0.8	2.992	4.212	6.355	8.842	4.003	5.691	8.613	11.948	4.729	6.719	10.080	13.826

Table 5. Effect of the power-law exponent and tapering parameter on the normalized flexural–torsional buckling load (P_{nor}) of simply supported thin-walled nanobeams with different nonlocal parameters (axial load applied on the TF at $x = l$).

μ (nm^2)	α	$m = 0.6$				$m = 1.2$				$m = 1.8$			
		$\beta = 0$	$\beta = 0.2$	$\beta = 0.5$	$\beta = 0.8$	$\beta = 0$	$\beta = 0.2$	$\beta = 0.5$	$\beta = 0.8$	$\beta = 0$	$\beta = 0.2$	$\beta = 0.5$	$\beta = 0.8$
0	0.0	3.881	5.299	7.808	10.770	5.218	7.133	10.513	14.484	6.160	8.392	12.304	16.869
	0.2	3.785	5.158	7.585	10.447	5.080	6.934	10.200	14.037	5.995	8.156	11.939	16.352
	0.5	3.648	4.961	7.277	10.005	4.889	6.658	9.775	13.434	5.766	7.830	11.441	15.652
	0.8	3.526	4.785	7.005	9.621	4.719	6.416	9.404	12.912	5.564	7.544	11.007	15.046
1.0	0.0	3.378	4.661	6.928	9.595	4.509	6.250	9.312	12.894	5.320	7.358	10.909	15.020
	0.2	3.288	4.531	6.726	9.308	4.382	6.066	9.030	12.498	5.168	7.141	10.581	14.566
	0.5	3.164	4.352	6.450	8.917	4.209	5.819	8.650	11.965	4.963	6.849	10.138	13.953
	0.8	3.054	4.195	6.209	8.577	4.060	5.605	8.323	11.507	4.787	6.598	9.756	13.423
3.0	0.0	2.659	3.748	5.668	7.908	3.465	4.968	7.585	10.607	4.038	5.832	8.891	12.354
	0.2	2.584	3.640	5.501	7.678	3.365	4.818	7.354	10.292	3.924	5.656	8.624	12.000
	0.5	2.484	3.494	5.276	7.364	3.235	4.621	7.047	9.868	3.779	5.429	8.270	11.519
	0.8	2.400	3.370	5.082	7.093	3.128	4.457	6.787	9.505	3.658	5.239	7.968	11.103

Under the first assumed load position, Figure 6 illustrates the variation in the normalized buckling load with respect to the web tapering ratio α and the flange tapering parameter β for different nonlocal parameters and material compositions, i.e., for a pure ceramic and AFG with $m = 1$. The same analysis is repeated for an axial load located at $x = L$, whose results are plotted in Figure 7, where the reduction in the buckling load is observable with increasing values of μ and decreased tapering ratios of α and β. Moreover, in Figures 8–10 we plot the lowest buckling load P_{nor} versus the tapering ratio for different values of gradient indexes, m, and nonlocal parameters ($\mu = 0$, 1, 2 and 3), while assuming $\alpha = \beta$, and three different positions of compressive load, as in the previous cases. Note that a beam subjected to a compressive axial force on the centroid can buckle in a pure torsional buckling mode. In this context, Table 6 addresses the influence of web and flange tapering ratios, material composition (power-law exponent), and nonlocal parameters ($\mu = 0$, 0.5 and 1) on the normalized torsional buckling load of the tapered nanobeam. Based on the results for both local and nonlocal beams and all non-uniformity ratios, the stability strength improves as the non-homogeneity parameter increases. In other words, a higher flexural–torsional buckling capacity is obtained with an increased power index, m, due to the increased content of ceramic phase and increased gradient index, under a fixed ratio of $\alpha = \beta$. As also visible in Figures 8–10, the rate of increase in the critical load with α, β is gradual for increased values of m. Based on a comparative evaluation of the results in Figures 8–10, with the same assumptions for α, β, m and μ, it seems that the load position has a significant effect on the stability strength of nanobeams with varying doubly symmetric I-sections, especially for higher values of the web tapering ratio. As also expected, the highest buckling capacity is obtained when the axial load is located exactly on the centroid, whereas the worst response corresponds to a compressive load applied on the TF owing to the compression of an initial bending moment resulting from a load eccentricity.

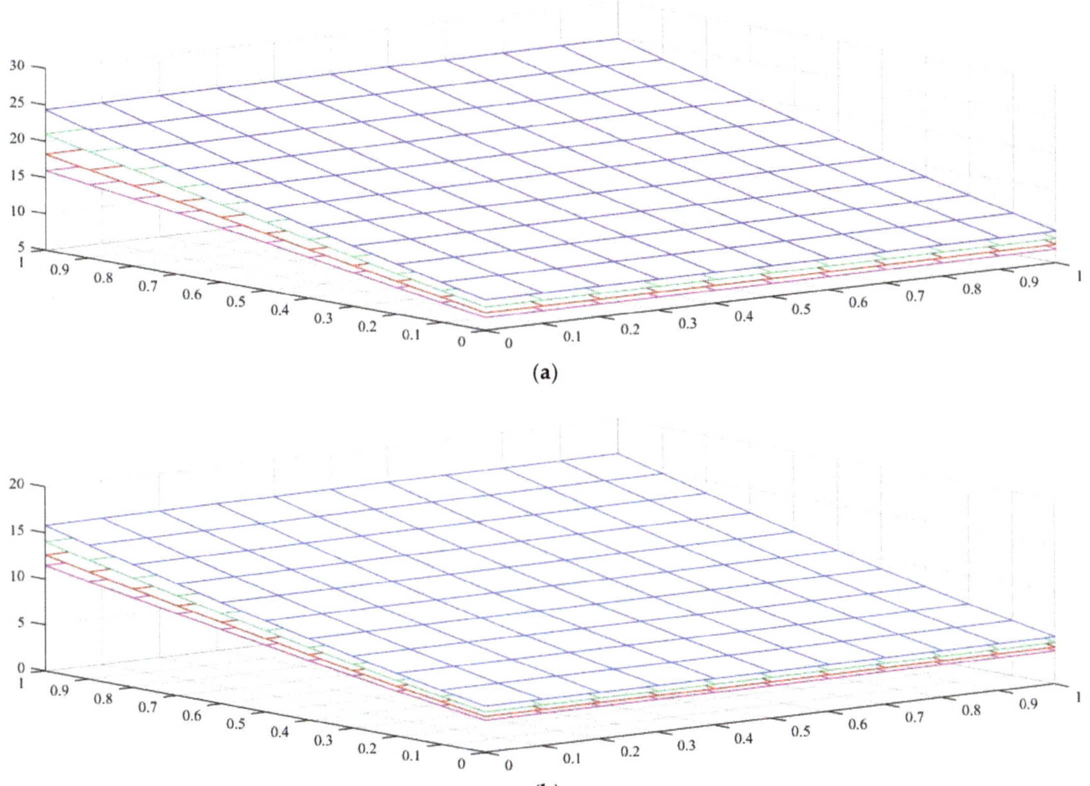

Figure 6. Variation in normalized buckling load (P_{nor}) of local and nonlocal beams with a tapered I-section for various webs. Effect of the power-law exponent and tapering parameter on the normalized torsional buckling load (P_{nor}) of simply supported thin-walled nanobeams with different nonlocal parameters (axial load applied on the Centroid): (**a**) pure ceramic, (**b**) AFG ($m = 1$).

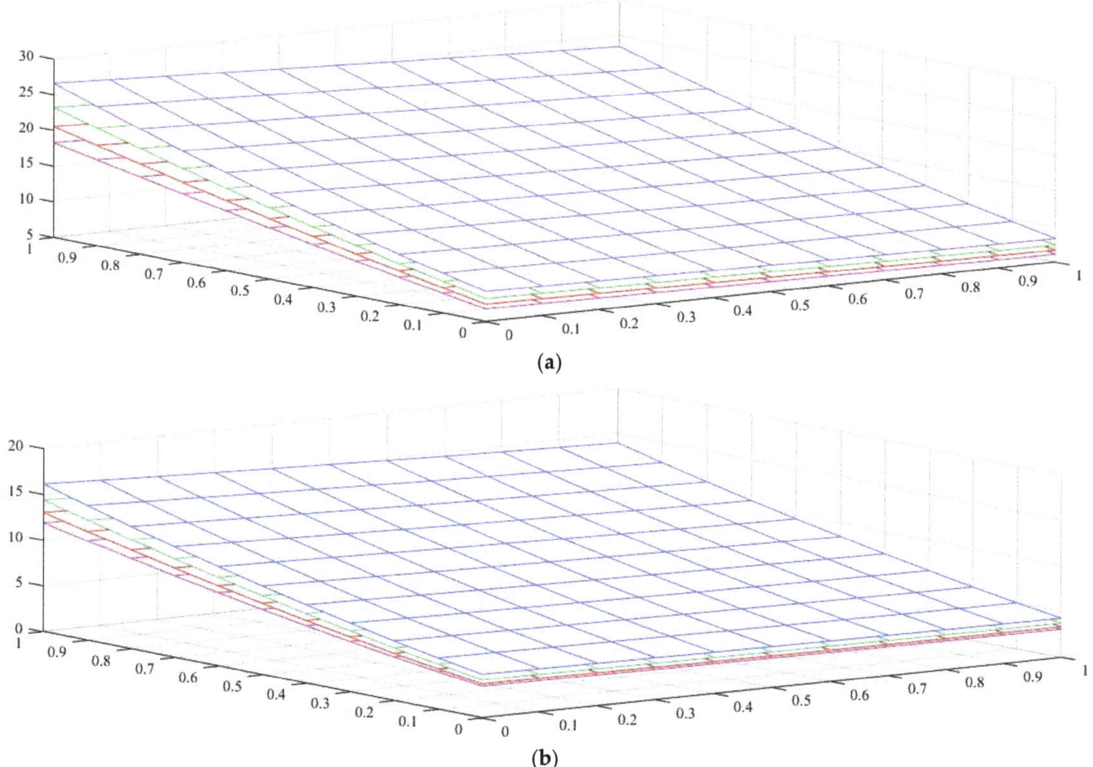

Figure 7. Variation in the normalized buckling load (P_{nor}) of local and nonlocal beams with a tapered I-section for various web and flange tapering ratios and four different Eringen's parameters (axial load on the TF at $x = L$): (**a**) pure ceramic, (**b**) AFG ($m = 1$).

Both tables and figures clearly show that the non-uniformity parameter has a remarkable influence on the flexural–torsional buckling load. For each selected power-law exponent, nonlocal parameter and loading position, the stability values of prismatic beams with $\alpha = \beta = 0$ and of double-tapered ones with $\alpha = \beta = 1$ are the lowest and highest, respectively. By consulting Tables 3–5 and Figures 6 and 7, one can also note that the response is much more affected by the flange tapering parameter β than the web non-uniformity ratio α, since the lowest flexural–torsional buckling modes usually occur with the lowest axis moment of inertia.

For beam-columns subjected to an axial load on the centroid and on the TF for $x = 0$, it is found that the buckling parameter increases with increasing values of both α and β, due to the enhancement of the cross-section's geometrical properties along with an overall increase in flexural and torsional stiffness in the elastic member (see Tables 3 and 4). The sensitivity of the mechanical response is slightly different for I-beams with an axial load applied on the TF at $x = L$. As shown in Table 5 and Figure 7, the linear stability strength of beams with a constant flange width tends to reduce with increasing values of α. This is mainly related to the fact that the initial bending moment M_y^* due to axial load eccentricity is enhanced by increasing the web tapering ratio α. The effect of this phenomenon on the buckling resistance of web and flange tapered beams is quite negligible when all the section walls have the same non-uniformity ratio (i.e., for $\alpha = \beta$).

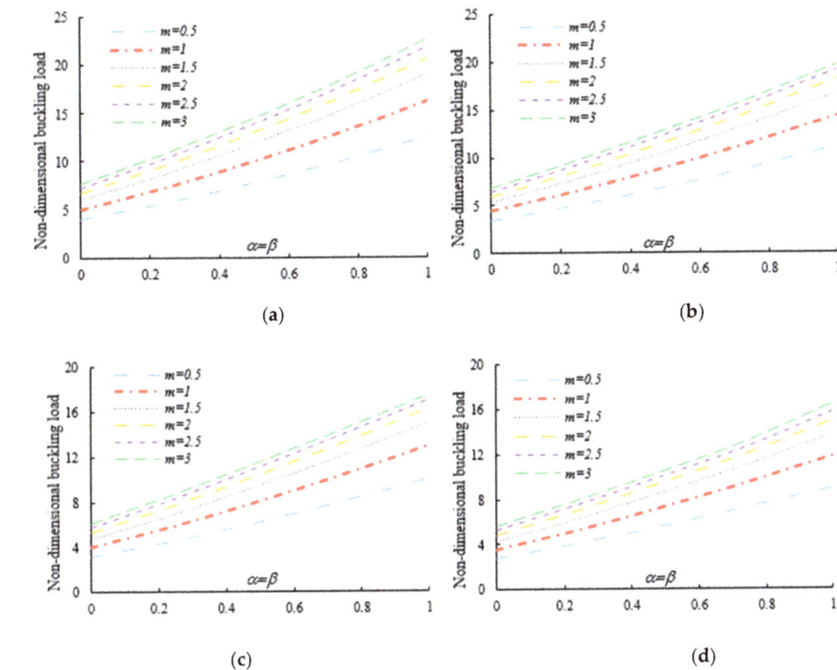

Figure 8. Variation in the flexural–torsional buckling load of I-tapered nanobeams with the tapering ratio and power-law exponent, for different nonlocality parameters (axial load on the TF at $x = 0$): (**a**) $\mu = 0$; (**b**) $\mu = 1$; (**c**) $\mu = 2$; (**d**) $\mu = 3$.

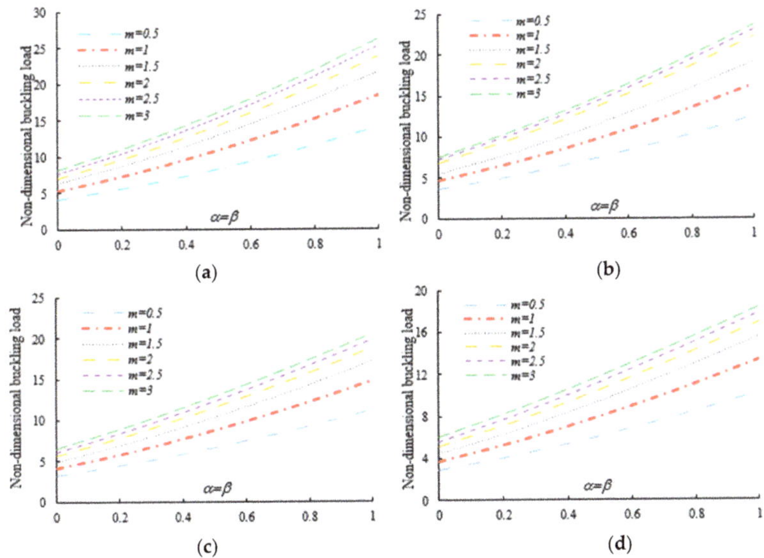

Figure 9. Variation in the flexural–torsional buckling load of I-tapered nanobeams with the tapering ratio and power-law exponent, for different nonlocality parameters (axial load on the centroid): (**a**) $\mu = 0$; (**b**) $\mu = 1$; (**c**) $\mu = 2$; (**d**) $\mu = 3$.

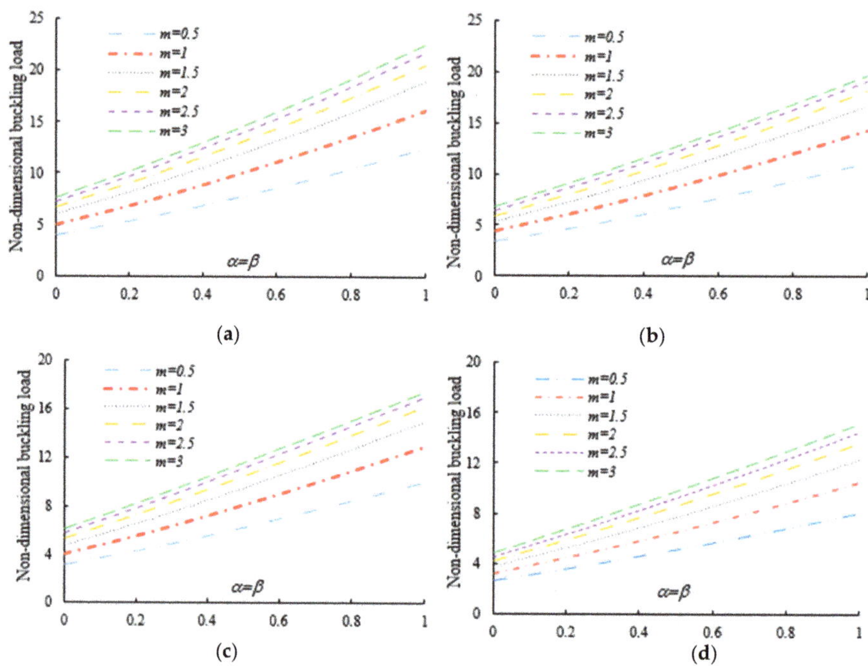

Figure 10. Variation in the flexural–torsional buckling load of I-tapered nanobeams with the tapering ratio and power-law exponent, for different nonlocality parameters (axial load on the TF at $x = L$): (**a**) $\mu = 0$; (**b**) $\mu = 1$; (**c**) $\mu = 2$; (**d**) $\mu = 3$.

Table 6. Effect of the power-law exponent and tapering parameter on the normalized torsional buckling load (P_{nor}) of simply supported thin-walled nanobeams with different nonlocal parameters (axial load applied on the Centroid).

	α	$\mu = 0$					$\mu = 0.5$					$\mu = 1.0$				
		$\beta = 0$	$\beta = 0.25$	$\beta = 0.5$	$\beta = 0.75$	$\beta = 1.0$	$\beta = 0$	$\beta = 0.25$	$\beta = 0.5$	$\beta = 0.75$	$\beta = 1.0$	$\beta = 0$	$\beta = 0.25$	$\beta = 0.5$	$\beta = 0.75$	$\beta = 1.0$
$m = 1$	0.0	26.399	31.400	36.266	41.068	45.851	23.116	28.587	33.610	38.395	43.059	19.151	25.731	31.080	35.936	40.536
	0.25	23.264	28.179	33.116	38.108	43.174	20.578	25.743	30.737	35.666	40.588	17.772	23.395	28.526	33.445	38.266
	0.5	21.048	25.814	30.706	35.739	40.919	18.770	23.667	28.555	33.493	38.514	16.526	21.651	26.582	31.464	36.362
	0.75	19.410	24.025	28.831	33.836	39.040	17.424	22.099	26.863	31.755	36.793	15.526	20.319	25.074	29.885	34.787
	1.0	18.155	22.633	27.343	32.291	37.476	16.386	20.881	25.526	30.351	35.365	14.726	19.276	23.882	28.610	33.484
$m = 2$	0.0	32.917	39.157	45.200	51.149	57.068	28.810	35.822	42.067	47.952	53.660	22.425	32.135	39.040	45.017	50.590
	0.25	29.212	35.391	41.552	47.747	54.009	25.880	32.480	38.722	44.811	50.846	21.761	29.544	36.068	42.145	48.015
	0.5	26.565	32.587	38.713	44.971	51.374	23.745	30.005	36.136	42.258	48.427	20.660	27.504	33.757	39.811	45.799
	0.75	24.596	30.445	36.477	42.709	49.145	22.138	28.118	34.109	40.187	46.387	19.612	25.915	31.943	37.923	43.935
	1.0	23.080	28.767	34.688	40.853	47.264	20.890	26.642	32.492	38.494	44.671	18.718	24.658	30.496	36.382	42.371
$m = 3$	0.0	36.991	43.765	50.277	56.664	63.005	32.637	40.321	46.990	53.227	59.259	24.922	36.356	43.842	50.097	55.888
	0.25	32.952	39.692	46.349	53.002	59.699	29.420	36.657	43.364	49.843	56.229	24.530	33.542	40.580	46.993	53.131
	0.5	30.034	36.619	43.249	49.973	56.819	27.038	33.908	40.519	47.052	53.592	23.509	31.257	38.008	44.432	50.724
	0.75	27.845	34.249	40.783	47.482	54.361	25.229	31.794	38.266	44.764	51.346	22.406	29.456	35.971	42.336	48.675
	1.0	26.151	32.379	38.794	45.421	52.272	23.816	30.130	36.454	42.879	49.441	21.420	28.021	34.335	40.613	46.942
Homogeneous	0.0	55.172	61.392	67.658	73.998	80.419	52.003	57.786	63.516	69.265	75.059	49.179	54.536	59.728	64.886	70.053
	0.25	49.161	55.583	62.099	68.739	75.511	46.296	52.418	58.496	64.612	70.798	43.724	49.607	55.278	60.896	66.524
	0.5	44.446	50.894	57.492	64.265	71.217	41.822	48.036	54.265	60.575	66.994	39.462	45.506	51.410	57.301	63.234
	0.75	40.765	47.145	53.724	60.521	67.539	38.353	44.530	50.784	57.168	63.700	36.189	42.219	48.196	54.216	60.321
	1.0	37.852	44.126	50.635	57.395	64.408	35.627	41.713	47.927	54.311	60.880	33.635	39.582	45.549	51.611	57.798

As was also expected, the nonlocal parameter shows a stiffness-softening effect and reduces the buckling strength for all the selected loading positions. Based on the plots in Figures 3–7, it seems that the effect of the Eringen's nonlocal parameter on the buckling response is more pronounced at higher tapering ratios and gradient indexes, especially for beams made of pure ceramic. For example, the normalized buckling load of prismatic members in Table 5 with $m = 1.2$ decreases by 36.5% when μ increases from 0 to 3. This can be explained by the fact that the flexural and torsional stiffness of simply supported tapered I-beams in a nonlocal theoretical context is inversely proportional to the Eringen's parameter. Usually, the introduction of a nonlocal effect increases the deflection response, or this increase is equivalent to the stiffness reduction in the structural member. Since the linear buckling resistance of beams is directly proportional to their stiffness, a meaningful decrease in the critical load is expected. In Figure 11, we represent the effect of the nonlocality parameter on the first four flexural–torsional buckling loads of nonlocal thin-walled beams with an I-section. In this way, we account for a compressive axial load applied at the TF for $x = 0$, while considering both an AFG prismatic beam with $m = 1$ and a homogeneous tapered beam with $\alpha = \beta = 0.5$. Based on the plots in Figure 11, it is clear that the nonlocal parameter has more pronounced effects on higher flexural–torsional buckling modes when compared with the lowest ones. It can also be stated that it is necessary to rely on nonlocal theories for an accurate estimation of the flexural–torsional stability limit of nanosized thin-walled beams at higher buckling modes. In addition, it is clearly observable that the nonlocal parameter effect increases when the μ ranges between 0 and 0.9.

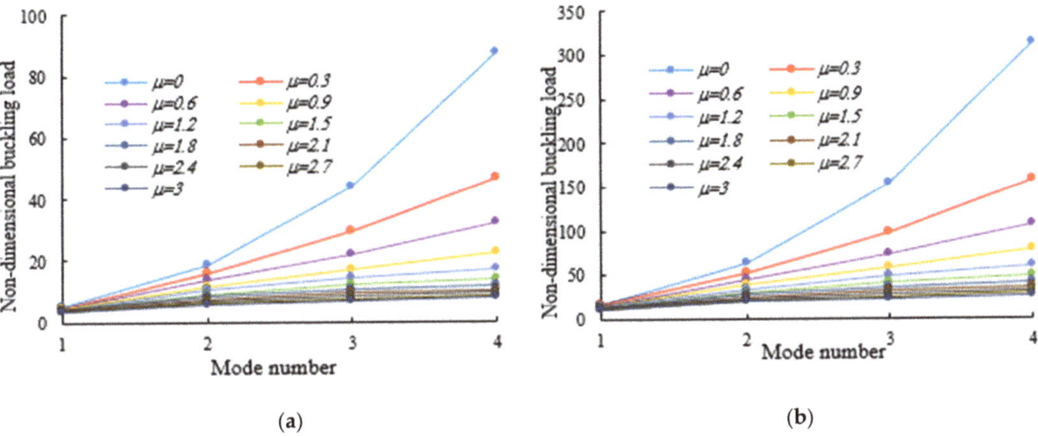

Figure 11. Effects of Eringen's nonlocal parameters on the flexural–torsional buckling load of doubly symmetric I-section nanobeams under a compressive axial load on the TF at x = 0: (**a**) $\alpha = \beta = 0$; $m = 1$; (**b**) $\alpha = \beta = 0.5$; Homogeneous.

5. Conclusions

In this paper, we explore the flexural–torsional buckling of AFG nanobeams with a varying I-section by resorting to the Vlasov model and Eringen's nonlocal elasticity theory. The material properties vary in the axial direction of structural elements according to the power-law distribution of the material constituents. The principle of minimum potential energy is applied to determine the governing equilibrium equations and boundary conditions for AFG tapered thin-walled nanobeams subjected to eccentric axial loads. The governing equations of the problem are implemented and solved numerically by means of the DQM in order to determine the flexural–torsional buckling load. A broad systematic investigation checks for the influence of some important parameters, including the web and flange tapering ratios, the nonlocal parameter, the mode number, the axial load eccentricity and the non-homogeneity index, on the overall response of doubly symmetric tapered nanobeams subjected to simply supported boundary conditions. For all the selected loading positions, it is found that the flexural–torsional buckling capacity of nanobeams with a tapered I-section decreases as the nonlocal parameter increases, whereas the buckling load increases as the flange tapering ratio and ceramic phase, Al_2O_3, increase. The effect of the flange tapering parameter β on the buckling capacity is more pronounced than that related to the web tapering ratio, α. As expected, the buckling capacity reaches its highest value in the absence of all possible eccentricity. In addition, the elastic buckling capacity of homogeneous double-tapered beams decreases as the nonlocal parameter increases, especially when compared to AFG prismatic I-beams. The small-scale effects become even more pronounced at higher buckling modes, such that they cannot be clearly disregarded when accurately defining the problem. In its present state, the formulation does not consider the grain sizes or the shapes of the alumina or aluminum components, but this will be considered in a more extended formulation that will include possible material anisotropies. A further extension of the proposed formulation will include the nonlinear effects on the coupled thermo-mechanical stability of tapered micro/nanosized systems, accounting for the possible presence of porosities and defects, together with different boundary and environmental conditions.

Author Contributions: Conceptualization, M.S., F.A., F.M., R.D. and F.T.; Formal analysis, M.S., F.A., F.M., R.D. and F.T.; Investigation, M.S., F.A. and F.M.; Validation, M.S., F.A., F.M., R.D. and F.T.; Writing—Original Draft, M.S., F.A. and F.M.; Writing—Review & Editing, R.D. and F.T. All authors have read and agreed to the published version of the manuscript.

Funding: This research received no external funding.

Conflicts of Interest: The authors declare no conflict of interest.

References

1. Yang, F.; Chong, A.; Lam, D.; Tong, P. Couple stress based strain gradient theory for elasticity. *Int. J. Solids Struct.* **2002**, *39*, 2731–2743. [CrossRef]
2. Gurtin, M.E.; Weissmüller, J.; Larche, F. A general theory of curved deformable interfaces in solids at equilibrium. *Philos. Mag. A* **1998**, *78*, 1093–1109. [CrossRef]
3. Eringen, A. Nonlocal polar elastic continua. *Int. J. Eng. Sci.* **1972**, *10*, 1–16. [CrossRef]
4. Eringen, A.C. On differential equations of nonlocal elasticity and solutions of screw dislocation and surface waves. *J. Appl. Phys.* **1983**, *54*, 4703–4710. [CrossRef]
5. Kitipornchai, S.; Trahair, N.S. Elastic behaviour of tapered monosymmetric I-beams. *J. Struct. Div.* **1975**, *101*, 1661–1678. [CrossRef]
6. Wekezer, J.W. Elastic torsion of thin walled bars of variable cross sections. *Comput. Struct.* **1984**, *19*, 401–407. [CrossRef]
7. Wekezer, J.W. Instability of thin-walled bars. *J. Eng. Mech. ASCE* **1985**, *111*, 923–935. [CrossRef]
8. Yang, Y.; Yau, J. Stability of Beams with Tapered I-Sections. *J. Eng. Mech.* **1987**, *113*, 1337–1357. [CrossRef]
9. Bradford, M.A.; Cuk, P.E. Elastic Buckling of Tapered Monosymmetric I-Beams. *J. Struct. Eng.* **1988**, *114*, 977–996. [CrossRef]
10. Baker, G. Lateral buckling of nonprismatic cantilevers using weighted residuals. *J. Eng. Mech.* **1993**, *119*, 1899–1919. [CrossRef]
11. Rajasekaran, S. Equations for Tapered Thin-Walled Beams of Generic Open Section. *J. Eng. Mech.* **1994**, *120*, 1607–1629. [CrossRef]
12. Rajasekaran, S. Instability of Tapered Thin?Walled Beams of Generic Section. *J. Eng. Mech.* **1994**, *120*, 1630–1640. [CrossRef]
13. Gupta, P.; Wang, S.T.; Blandford, G.E. Lateral-Torsional Buckling of Nonprismatic I-Beams. *J. Struct. Eng.* **1996**, *122*, 748–755. [CrossRef]
14. Ronagh, H.; Bradford, M.; Attard, M. Nonlinear analysis of thin-walled members of variable cross-section. Part I: Theory. *Comput. Struct.* **2000**, *77*, 285–299. [CrossRef]
15. Chen, C.-N. Dynamic Equilibrium Equations of Non-Prismatic Beams Defined on an Arbitrarily Selected Co-Òrdinate System. *J. Sound Vib.* **2000**, *230*, 241–260. [CrossRef]
16. Kim, S.-B.; Kim, M.-Y. Improved formulation for spatial stability and free vibration of thin-walled tapered beams and space frames. *Eng. Struct.* **2000**, *22*, 446–458. [CrossRef]
17. Li, G.-Q.; Li, J.-J. A tapered Timoshenko–Euler beam element for analysis of steel portal frames. *J. Constr. Steel Res.* **2002**, *58*, 1531–1544. [CrossRef]
18. Peddieson, J.; Buchanan, G.R.; McNitt, R.P. Application of nonlocal continuum models to nanotechnology. *Int. J. Eng. Sci.* **2003**, *41*, 305–312. [CrossRef]
19. Elishakoff, I.; Guédé, Z. Analytical Polynomial Solutions for Vibrating Axially Graded Beams. *Mech. Adv. Mater. Struct.* **2004**, *11*, 517–533. [CrossRef]

20. Andrade, A.; Camotim, D. Lateral–Torsional Buckling of Singly Symmetric Tapered Beams: Theory and Applications. *J. Eng. Mech.* **2005**, *131*, 586–597. [CrossRef]
21. Andrade, A.; Camotim, D.; Dinis, P.B. Lateral-torsional buckling of singly symmetric web-tapered thin-walled I-beams: 1D model vs. shell FEA. *Comput. Struct.* **2007**, *85*, 1343–1359. [CrossRef]
22. Andrade, A.; Camotim, D.; e Costa, P.P. On the evaluation of elastic critical moments in doubly and singly symmetric I-section cantilevers. *J. Constr. Steel Res.* **2007**, *63*, 894–908. [CrossRef]
23. Mohri, F.; Azrar, L.; Potier-Ferry, M. Lateral post-buckling analysis of thin-walled open section beams. *Thin-Walled Struct.* **2002**, *40*, 1013–1036. [CrossRef]
24. Mohri, F.; Damil, N.; Potier-Ferry, M.; Mohri, F.; Damil, N.; Potier-Ferry, M. Buckling and lateral buckling interaction in thin-walled beam-column elements with mono-symmetric cross sections. *Appl. Math. Model.* **2013**, *37*, 3526–3540. [CrossRef]
25. Samanta, A.; Kumar, A. Distortional buckling in monosymmetric I-beams. *Thin-Walled Struct.* **2006**, *44*, 51–56. [CrossRef]
26. Reddy, J. Nonlocal theories for bending, buckling and vibration of beams. *Int. J. Eng. Sci.* **2007**, *45*, 288–307. [CrossRef]
27. Challamel, N.; Andrade, A.; Camotim, D. An Analytical Study on The Lateral-Torsional Buckling of Linearly Tapered Cantilever Strip Beams. *Int. J. Struct. Stab. Dyn.* **2007**, *7*, 441–456. [CrossRef]
28. Wang, C.M.; Zhang, Y.Y.; He, X.Q. Vibration of Non-local Timoshenko Beams. *Nanotechnology* **2007**, *18*, 105401. [CrossRef]
29. Pradhan, S.; Sarkar, A. Analyses of tapered fgm beams with nonlocal theory. *Struct. Eng. Mech.* **2009**, *32*, 811–833. [CrossRef]
30. Aydogdu, M. A general nonlocal beam theory: Its application to nanobeam bending, buckling and vibration. *Phys. E Low-Dimens. Syst. Nanostruct.* **2009**, *41*, 1651–1655. [CrossRef]
31. Civalek, Ö.; Akgöz, B. Free vibration analysis of microtubules as cytoskeleton components: Nonlocal Euler-Bernoulli beam modeling. *Sci. Iran. Trans. B Mech. Eng.* **2010**, *17*, 367–375.
32. Danesh, M.; Farajpour, A.; Mohammadi, M. Axial vibration analysis of a tapered nanorod based on nonlocal elasticity theory and differential quadrature method. *Mech. Res. Commun.* **2012**, *39*, 23–27. [CrossRef]
33. Şimşek, M.; Yurtcu, H. Analytical solutions for bending and buckling of functionally graded nanobeams based on the nonlocal Timoshenko beam theory. *Compos. Struct.* **2013**, *97*, 378–386. [CrossRef]
34. McCann, F.; Gardner, L.; Wadee, M.A. Design of steel beams with discrete lateral restraints. *J. Constr. Steel Res.* **2013**, *80*, 82–90. [CrossRef]
35. Eltaher, M.; Alshorbagy, A.; Mahmoud, F. Determination of neutral axis position and its effect on natural frequencies of functionally graded macro/nanobeams. *Compos. Struct.* **2013**, *99*, 193–201. [CrossRef]
36. Eltaher, M.; Emam, S.A.; Mahmoud, F. Static and stability analysis of nonlocal functionally graded nanobeams. *Compos. Struct.* **2013**, *96*, 82–88. [CrossRef]
37. Shahba, A.; Attarnejad, R.; Hajilar, S. A Mechanical-Based Solution for Axially Functionally Graded Tapered Euler-Bernoulli Beams. *Mech. Adv. Mater. Struct.* **2013**, *20*, 696–707. [CrossRef]
38. Benyamina, A.B.; Meftah, S.A.; Mohri, F.; Daya, E.M. Analytical solutions attempt for lateral torsional buckling of doubly symmetric web-tapered I-beams. *Eng. Struct.* **2013**, *56*, 1207–1219. [CrossRef]
39. Nguyen, C.T.; Moon, J.; Le, V.N.; Lee, H.-E. Lateral–torsional buckling of I-girders with discrete torsional bracings. *J. Constr. Steel Res.* **2010**, *66*, 170–177. [CrossRef]
40. Attard, M.; Kim, M.-Y. Lateral buckling of beams with shear deformations—A hyperelastic formulation. *Int. J. Solids Struct.* **2010**, *47*, 2825–2840. [CrossRef]
41. Challamel, N.; Wang, C.M. Exact lateral–torsional buckling solutions for cantilevered beams subjected to intermediate and end transverse point loads. *Thin-Walled Struct.* **2010**, *48*, 71–76. [CrossRef]
42. Ke, L.-L.; Wang, Y.-S. Size effect on dynamic stability of functionally graded microbeams based on a modified couple stress theory. *Compos. Struct.* **2011**, *93*, 342–350. [CrossRef]
43. De Borbón, F.; Mirasso, A.; Ambrosini, D. A beam element for coupled torsional-flexural vibration of doubly unsymmetrical thin walled beams axially loaded. *Comput. Struct.* **2011**, *89*, 1406–1416. [CrossRef]
44. Serna, M.A.; Ibáñez, J.R.; López, A. Elastic flexural buckling of non-uniform members: Closed-form expression and equivalent load approach. *J. Constr. Steel Res.* **2011**, *67*, 1078–1085. [CrossRef]
45. Akgöz, B.; Civalek, Ö. Free vibration analysis of axially functionally graded tapered Bernoulli–Euler microbeams based on the modified couple stress theory. *Compos. Struct.* **2013**, *98*, 314–322. [CrossRef]
46. Lezgy-Nazargah, M.; Vidal, P.; Polit, O. An efficient finite element model for static and dynamic analyses of functionally graded piezoelectric beams. *Compos. Struct.* **2013**, *104*, 71–84. [CrossRef]
47. Trahair, N. Bending and buckling of tapered steel beam structures. *Eng. Struct.* **2014**, *59*, 229–237. [CrossRef]
48. Ke, L.-L.; Wang, Y.-S. Free vibration of size-dependent magneto-electro-elastic nanobeams based on the nonlocal theory. *Phys. E Low-Dimens. Syst. Nanostruct.* **2014**, *63*, 52–61. [CrossRef]
49. Rahmani, O.; Pedram, O. Analysis and modeling the size effect on vibration of functionally graded nanobeams based on nonlocal Timoshenko beam theory. *Int. J. Eng. Sci.* **2014**, *77*, 55–70. [CrossRef]
50. Liu, C.; Ke, L.L.; Wang, Y.S.; Yang, J.; Kitipornchai, S. Buckling and post-buckling of size-dependent piezoelectric Timoshenko nanobeams subject to thermo-electro-mechanical loadings. *Int. J. Struct. Stab. Dyn.* **2014**, *14*. [CrossRef]
51. Nami, M.R.; Janghorban, M.; Damadam, M. Thermal buckling analysis of functionally graded rectangular nanoplates based on nonlocal third-order shear deformation theory. *Aerosp. Sci. Technol.* **2015**, *41*, 7–15. [CrossRef]

52. Mohri, F.; Meftah, S.A.; Damil, N. A large torsion beam finite element model for tapered thin-walled open cross sections beams. *Eng. Struct.* **2015**, *99*, 132–148. [CrossRef]

53. Kus, J. Lateral-torsional buckling steel beams with simultaneously tapered flanges and web. *Steel Compos. Struct.* **2015**, *19*, 897–916. [CrossRef]

54. Pandeya, A.; Singhb, J. A variational principle approach for vibration of non-uniform nanocantilever using nonlocal elasticity theory. *Proced. Mater. Sci.* **2015**, *10*, 497–506. [CrossRef]

55. Shafiei, N.; Kazemi, M.; Safi, M.; Ghadiri, M. Nonlinear vibration of axially functionally graded non-uniform nanobeams. *Int. J. Eng. Sci.* **2016**, *106*, 77–94. [CrossRef]

56. Zhang, Y.; He, B.; Yao, L.-K.; Long, J. Buckling analysis of thin-walled members via semi-analytical finite strip transfer matrix method. *Adv. Mech. Eng.* **2016**, *8*, 1–11. [CrossRef]

57. Zhang, W.-F.; Liu, Y.-C.; Hou, G.-L.; Chen, K.-S.; Ji, J.; Deng, Y.; Deng, S.-L. Lateral-torsional buckling analysis of cantilever beam with tip lateral elastic brace under uniform and concentrated load. *Int. J. Steel Struct.* **2016**, *16*, 1161–1173. [CrossRef]

58. Kiani, K. Free dynamic analysis of functionally graded tapered nanorods via a newly developed nonlocal surface energy-based integro-differential model. *Compos. Struct.* **2016**, *139*, 151–166. [CrossRef]

59. Mohammadimehr, M.; Mohandes, M.; Moradi, M. Size dependent effect on the buckling and vibration analysis of double-bonded nanocomposite piezoelectric plate reinforced by boron nitride nanotube based on modified couple stress theory. *J. Vib. Control* **2016**, *22*, 1790–1807. [CrossRef]

60. Arefi, M.; Zenkour, A.M. Employing sinusoidal shear deformation plate theory for transient analysis of three layers sandwich nanoplate integrated with piezo-magnetic face-sheets. *Smart Mater. Struct.* **2016**, *25*, 115040. [CrossRef]

61. Bourihane, O.; Ed-Dinari, A.; Braikat, B.; Jamal, M.; Mohri, F.; Damil, N. Stability analysis of thin-walled beams with open section subject to arbitrary loads. *Thin-Walled Struct.* **2016**, *105*, 156–171. [CrossRef]

62. Ebrahimi, F.; Salari, E. Thermo-mechanical vibration analysis of nonlocal temperature-dependent FG nanobeams with various boundary conditions. *Compos. Part B Eng.* **2015**, *78*, 272–290. [CrossRef]

63. Ebrahimi, F.; Barati, M.R. A Nonlocal Higher-Order Shear Deformation Beam Theory for Vibration Analysis of Size-Dependent Functionally Graded Nanobeams. *Arab. J. Sci. Eng.* **2016**, *41*, 1679–1690. [CrossRef]

64. Ebrahimi, F.; Barati, M.R. Vibration analysis of nonlocal beams made of functionally graded material in thermal environment. *Eur. Phys. J. Plus* **2016**, *131*. [CrossRef]

65. Ebrahimi, F.; Shaghaghi, G.R.; Boreiry, M. A Semi-Analytical Evaluation of Surface and Nonlocal Effects on Buckling and Vibrational Characteristics of Nanotubes with Various Boundary Conditions. *Int. J. Struct. Stab. Dyn.* **2016**, *16*. [CrossRef]

66. Ebrahimi, F.; Barati, M.R. Buckling analysis of nonlocal third-order shear deformable functionally graded piezoelectric nanobeams embedded in elastic medium. *J. Braz. Soc. Mech. Sci. Eng.* **2017**, *39*, 937–952. [CrossRef]

67. Ebrahimi, F.; Barati, M.R. A nonlocal strain gradient refined beam model for buckling analysis of size-dependent shear-deformable curved FG nanobeams. *Compos. Struct.* **2017**, *159*, 174–182. [CrossRef]

68. Maalawi, K.Y. Dynamic optimization of functionally graded thin-walled box beams. *Int. J. Struct. Stab. Dyn.* **2017**, *17*, 1750109. [CrossRef]

69. Lezgy-Nazargah, M. A generalized layered global-local beam theory for elasto-plastic analysis of thin-walled members. *Thin-Walled Struct.* **2017**, *115*, 48–57. [CrossRef]

70. Nguyen, T.-T.; Thang, P.T.; Lee, J. Flexural-torsional stability of thin-walled functionally graded open-section beams. *Thin-Walled Struct.* **2017**, *110*, 88–96. [CrossRef]

71. Nguyen, T.-T.; Thang, P.T.; Lee, J. Lateral buckling analysis of thin-walled functionally graded open-section beams. *Compos. Struct.* **2017**, *160*, 952–963. [CrossRef]

72. Li, X.; Li, L.; Hu, Y.; Ding, Z.; Deng, W. Bending, buckling and vibration of axially functionally graded beams based on nonlocal strain gradient theory. *Compos. Struct.* **2017**, *165*, 250–265. [CrossRef]

73. Khaniki, H.B.; Hashemi, S.H. Dynamic transverse vibration characteristics of nonuniform nonlocal strain gradient beams using the generalized differential quadrature method. *Eur. Phys. J. Plus* **2017**, *132*, 500. [CrossRef]

74. Hashemi, S.H.; Khaniki, H.B. Dynamic Behavior of Multi-Layered Viscoelastic Nanobeam System Embedded in a Viscoelastic Medium with a Moving Nanoparticle. *J. Mech.* **2017**, *33*, 559–575. [CrossRef]

75. Khaniki, H.B.; Hosseini-Hashemi, S. Buckling analysis of tapered nanobeams using nonlocal strain gradient theory and a generalized differential quadrature method. *Mater. Res. Express* **2017**, *4*, 065003. [CrossRef]

76. Rajasekaran, S.; Khaniki, H.B. Bending, buckling and vibration of small-scale tapered beams. *Int. J. Eng. Sci.* **2017**, *120*, 172–188. [CrossRef]

77. Khaniki, H.B. On vibrations of nanobeam systems. *Int. J. Eng. Sci.* **2018**, *124*, 85–103. [CrossRef]

78. Khaniki, H.B. Vibration analysis of rotating nanobeam systems using Eringen's two-phase local/nonlocal model. *Physica E* **2018**, *99*, 310–319. [CrossRef]

79. Rajasekaran, S.; Khaniki, H.B. Finite element static and dynamic analysis of axially functionally graded nonuniform small-scale beams based on nonlocal strain gradient theory. *Mech. Adv. Mater. Struct.* **2018**, *26*, 1245–1259. [CrossRef]

80. Khaniki, H.B.; Rajasekaran, S. Mechanical analysis of non-uniform bi-directional functionally graded intelligent micro-beams using modified couple stress theory. *Mater. Res. Express* **2018**, *5*, 055703. [CrossRef]

81. Koutoati, K.; Mohri, F.; Daya, E.M. Finite element approach of axial bending coupling on static and vibration behaviors of functionally graded material sandwich beams. *Mech. Adv. Mater. Struct.* **2021**, *28*, 1537–1553. [CrossRef]
82. Glabisz, W.; Jarczewska, K.; Hołubowski, R. Stability of Timoshenko beams with frequency and initial stress dependent nonlocal parameters. *Arch. Civ. Mech. Eng.* **2019**, *19*, 1116–1126. [CrossRef]
83. Ebrahimi, F.; Barati, M.R.; Civalek, Ö. Application of Chebyshev–Ritz method for static stability and vibration analysis of nonlocal microstructure-dependent nanostructures. *Eng. Comput.* **2020**, *36*, 953–964. [CrossRef]
84. Jrad, W.; Mohri, F.; Robin, G.; Daya, E.M.; Al-Hajjar, J. Analytical and finite element solutions of free and forced vibration of unrestrained and braced thin-walled beams. *J. Vib. Control.* **2019**, *26*, 255–276. [CrossRef]
85. Arefi, M.; Civalek, O. Static analysis of functionally graded composite shells on elastic foundations with nonlocal elasticity theory. *Arch. Civ. Mech. Eng.* **2020**, *20*, 1–17. [CrossRef]
86. Osmani, A.; Meftah, S.A. Lateral buckling of tapered thin walled bi-symmetric beams under combined axial and bending loads with shear deformations allowed. *Eng. Struct.* **2018**, *165*, 76–87. [CrossRef]
87. Chen, Z.; Li, J.; Sun, L. Calculation of Critical Lateral-Torsional Buckling Loads of Beams Subjected to Arbitrarily Transverse Loads. *Int. J. Struct. Stab. Dyn.* **2019**, *19*. [CrossRef]
88. Achref, H.; Foudil, M.; Cherif, B. Higher buckling and lateral buckling strength of unrestrained and braced thin-walled beams: Analytical, numerical and design approach applications. *J. Constr. Steel Res.* **2019**, *155*, 1–19. [CrossRef]
89. Asgarian, B.; Soltani, M.; Mohri, F.; Asgarian, B.; Soltani, M.; Mohri, F. Lateral-torsional buckling of tapered thin-walled beams with arbitrary cross-sections. *Thin-Walled Struct.* **2013**, *62*, 96–108. [CrossRef]
90. Soltani, M.; Gharebaghi, S.A.; Mohri, F. Lateral stability analysis of steel tapered thin-walled beams under various boundary conditions. *Numer. Methods Civ. Eng.* **2018**, *3*, 13–25. [CrossRef]
91. Soltani, M.; Asgarian, B.; Mohri, F. Improved finite element formulation for lateral stability analysis of axially functionally graded non-prismatic I-beams. *Int. J. Struct. Stab. Dyn.* **2019**, *19*, 1950108. [CrossRef]
92. Lal, A.; Markad, K. Thermo-Mechanical Post Buckling Analysis of Multiwall Carbon Nanotube-Reinforced Composite Laminated Beam under Elastic Foundation. *Curved Layer. Struct.* **2019**, *6*, 212–228. [CrossRef]
93. Pavlović, I.R.; Pavlović, R.; Janevski, G.; Despenić, N.; Pajković, V. Dynamic Behavior of Two Elastically Connected Nanobeams Under a White Noise Process. *Facta Univ. Ser. Mech. Eng.* **2020**, *18*, 219–227. [CrossRef]
94. Soltani, M.; Atoufi, F.; Mohri, F.; Dimitri, R.; Tornabene, F. Nonlocal elasticity theory for lateral stability analysis of tapered thin-walled nanobeams with axially varying materials. *Thin-Walled Struct.* **2021**, *159*, 107268. [CrossRef]
95. Vlasov, V.Z. *Thin-Walled Elastic Beams, Moscow, 1959 (French Translation, Pièces Longues en Voiles Minces)*; Eyrolles: Paris, France, 1962.
96. Bellman, R.; Casti, J. Differential quadrature and long-term integration. *J. Math. Anal. Appl.* **1971**, *34*, 235–238. [CrossRef]
97. Bert, C.W.; Malik, M. Differential Quadrature Method in Computational Mechanics: A Review. *Appl. Mech. Rev.* **1996**, *49*, 1–28. [CrossRef]
98. Shu, C. *Differential Quadrature and Its Application in Engineering*; Springer Science and Business Media LLC: Berlin, Germany, 2000.
99. Zong, Z.; Zhang, Y. Advanced Differential Quadrature Methods. Chapman & Hall/CRC: Boca Raton, FL, USA, 2009.
100. Tornabene, F.; Dimitri, R.; Viola, E. Transient dynamic response of generally-shaped arches based on a GDQ-time-stepping method. *Int. J. Mech. Sci.* **2016**, *114*, 277–314. [CrossRef]
101. Dimitri, R.; Fantuzzi, N.; Tornabene, F.; Zavarise, G. Innovative numerical methods based on SFEM and IGA for computing stress concentrations in isotropic plates with discontinuities. *Int. J. Mech. Sci.* **2016**, *118*, 166–187. [CrossRef]
102. Tornabene, F.; Brischetto, S. 3D capability of refined GDQ models for the bending analysis of composite and sandwich plates, spherical and doubly-curved shells. *Thin-Walled Struct.* **2018**, *129*, 94–124. [CrossRef]
103. Fazzolari, F.A.; Viscoti, M.; Dimitri, R.; Tornabene, F. 1D-Hierarchical Ritz and 2D-GDQ Formulations for the free vibration analysis of circular/elliptical cylindrical shells and beam structures. *Compos. Struct.* **2021**, *258*, 113338. [CrossRef]
104. *ANSYS, Version 5.4*; Swanson Analysis System, Inc.: Canonsburg, PA, USA, 2007.

 *nanomaterials*

MDPI

Article

Computational Study on Surface Bonding Based on Nanocone Arrays

Xiaohui Song [1,2], Shunli Wu [2] and Rui Zhang [1,*]

1   School of Mechanical Engineering, Zhengzhou University, Zhengzhou 450001, China; xhsong@foxmail.com
2   Institute of Applied Physics, Henan Academy of Sciences, Zhengzhou 450008, China; yukephysics@gmail.com
*   Correspondence: lyzr@zzu.edu.cn

Abstract: Surface bonding is an essential step in device manufacturing and assembly, providing mechanical support, heat transfer, and electrical integration. Molecular dynamics simulations of surface bonding and debonding failure of copper nanocones are conducted to investigate the underlying adhesive mechanism of nanocones and the effects of separation distance, contact length, temperature, and size of the cones. It is found that van der Waals interactions and surface atom diffusion simultaneously contribute to bonding strength, and different adhesive mechanisms play a main role in different regimes. The results reveal that increasing contact length and decreasing separation distance can simultaneously contribute to increasing bonding strength. Furthermore, our simulations indicate that a higher temperature promotes diffusion across the interface so that subsequent cooling results in better adhesion when compared with cold bonding at the same lower temperature. It also reveals that maximum bonding strength was obtained when the cone angle was around 53°. These findings are useful in designing advanced metallic bonding processes at low temperatures and pressure with tenable performance.

Keywords: surface bonding; nanocone arrays; molecular dynamics simulation

Citation: Song, X.; Wu, S.; Zhang, R. Computational Study on Surface Bonding Based on Nanocone Arrays. *Nanomaterials* **2021**, *11*, 1369. https://doi.org/10.3390/nano11061369

Academic Editor: Francisco Torrens

Received: 22 April 2021
Accepted: 19 May 2021
Published: 21 May 2021

Publisher's Note: MDPI stays neutral with regard to jurisdictional claims in published maps and institutional affiliations.

1. Introduction

Surface bonding is an essential step in device manufacturing and assembly, providing mechanical support, heat transfer, and electrical integration. Traditional surface bonding techniques in electronic assembly strongly rely on high-temperature processes such as reflow soldering, which can lead to undesirable thermal damage, toxic solder materials pollution, and residual stress at the bonding interface [1,2]. Besides, thermo-compression is another widely used bonding approach because it provides intrinsic interconnections and excellent bonding strength. However, high bonding pressure and temperature may result in bond-alignment deviation, high thermal stress, and possible damages to bonded devices [3].

Consequently, to achieve high-density integration, high performance, and low power consumption, new high-level surface bonding schemes have to be developed. Recently, the process compatibility and reliability of cold bonding between surfaces with patterned arrangements of nanowires, nanoparticles, and nanocones have been improved by studies of the effects of lowering the temperature and pressure for bonding [4–8]. Such nanometal bonding methods exhibit high bonding strength and low electrical resistance at the interface at ambient or low temperatures. A maximum adhesion strength of $16.4 \, \text{N/cm}^2$ was achieved using a bent, hook-like nanowire surface fastener, showing a room temperature bonding ability and adaptability to a highly ordered electrode such as the ball grid array (BGA) [9]. In addition, a new copper/polystyrene core/shell nanowire surface fastener showed a higher adhesion strength ($\sim 44.42 \, \text{N/cm}^2$) and a much lower contact resistivity ($\sim 0.75 \times 10^{-2} \, \Omega \cdot \text{cm}^2$) [10]. Furthermore, the metallic nano structures exhibit good compatibility with a bendable or stretchable device, and can be fabricated on flexible electronic circuits, making them attractive as adhesive materials for wearable applications [11].

To design a nanometal structure with optimum properties, it is crucial to understand the atomic bonding mechanisms at the nanoscale and accordingly optimize the bonding process. Several mechanisms and theoretical models, such as van der Waals (vdW) interactions [12], surface coalescence [13,14], and Amonton's first law [15], have been proposed to describe the processes of metal nanowires and nanoparticles-based surface bonding. However, some novel phenomena observed in nanometal-based surface bonding suggest that different nanostructures relate to different bonding mechanisms. All too often, a single mechanism is described as the cause of surface bonding when nanometal bonding may be a combination of mechanisms that contribute to the nanoscale adhesion. It was reported previously that metal nanocones-based surface bonding, which has relative controllability and operational fabrication processes, can be performed at room temperature [16,17]. However, to our knowledge, the discussion of metal nanocones-based surface bonding from the micro mechanism is still rare.

In this work, we explored the atomic interactions that lead to the formation of surface joints between copper nanocones using molecular dynamics (MD) simulations. Conical nanometals with various diameters and heights were then placed in close proximity at different separation distances and contact lengths to form nanoscale surface bonding at the interface. The debonding process and the effect of the temperature on the mechanical properties of the interface bonding between copper nanocones were investigated.

2. Computational Methods

Molecular dynamics (MD) simulation has been proven to be a powerful tool at the nanoscale and is widely used in the studies of nanomaterials [18]. In the present study, MD simulations were used to deal with the bonding and debonding process of the copper nanocones interface joint. Our simulations were based on the massively parallel LAMMPS code [19]. The visualization was based on Open Visualization Tool (OVITO) [20]. In this work, we modeled copper nanocones with various diameters and heights. As shown in Figure 1, the contact length represents the overlap depth of two cones, the separation distance represents the distance between two cone axes. This study involves a combination of embedded atom method (EAM) potential and Lennard-Jones potential [21,22]. The EAM potential [23–25] is proven to accurately depict the many-body atomic interactions in metallic systems, and is widely used to simulate the deformation behavior, indentation behavior of thin copper, and Cu composites [26–28]. Therefore, the EAM potential was used to calculate the interatomic interactions between copper atoms inside the nanocone; while, in the simulation of the contact process, the Lennard-Jones (LJ) pair potential, which involves nonbond long-range interactions, was used to describe the interactions between copper atoms located at the interface of the adjacent nanocones on the opposite surfaces of the joint. For fcc Cu, an LJ potential with parameters σ = 2.228 Å, ε = 0.415 eV and a cutoff radius = 4 σ was used in the simulation [29,30].

The initial copper nanocone structure with a diameter of 80 Å and height of 100 Å was cut from perfect bulk face-centered cubic (fcc) crystals with lattice constant of 0.3615 nm by atomsk [31]. A single cone was equilibrated and optimized using the conjugate gradient algorithm to perform an energy minimization of the system for 300 ps, with a time step of 0.5 fs afterwards. Additionally, the two nanocones formed a surface joint with a contact length of 20 Å and a separation distance of 20 Å, as shown in Figure 1, which allowed the two cones to be attracted to each other and form cold bonds under diffusion and intermixing of surface atoms. The configuration contained 29,192 atoms. As shown in Figure 1, the separation distance represents the distance between the parallel axes of the two cones, and the contact length represents overlapping depth of the two cones. During simulation, non-periodic boundary conditions were implemented in each direction. We conducted a surface bonding simulation with the displaced layer and the fixed layer was fixed. As shown in Figure 2, the simulation model underwent an equilibration process for

300 ps in the NVT ensemble at a constant temperature using a Nose–Hoover thermostat with a simulation input temperature of 300 K [32].

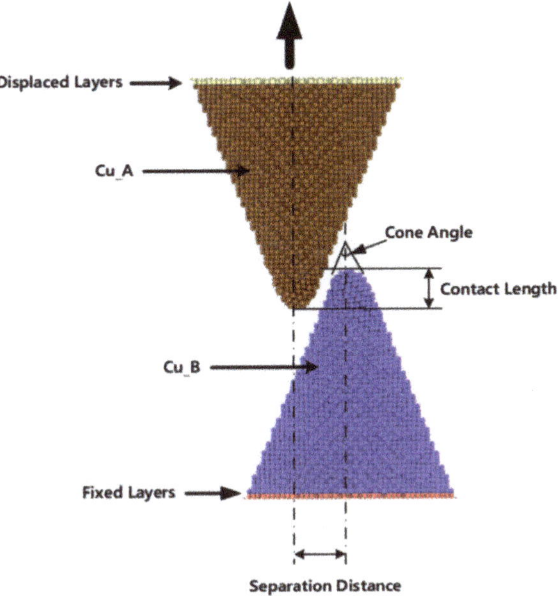

Figure 1. Copper nanocones surface bonding simulation model.

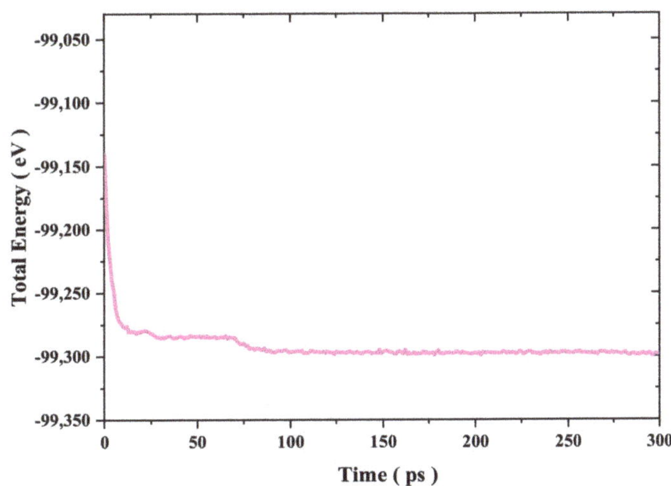

Figure 2. Total energy variation during 300 ps equilibration of a pair of nanocones, as oriented in Figure 1, with an initial axial separation of 20 Å and a contact length of 20 Å.

To analyze the mechanical properties, such as the maximum load that can be sustained by the formed bonding joint, we performed MD simulations for the debonding process under tensile loading. First, the fixed layer was fixed while the displaced layer displaced along the cone axis by 0.1 Å. Second, the strained structure was equilibrated at the constant temperature for 30 ps in the NVT ensemble. The loading force acting on the fixed layer is averaged over the last 5 ps to reduce fluctuations. This process was repeated until the structure fractures.

3. Results and Discussion

As shown in Figure 3, the configuration modification of two adjacent copper nanocones during the bonding and the debonding process was investigated. In the initial state, two nanocones formed cold bonds and surface joints at the interface with a certain contact length and separation distance at a temperature of 300 K after equilibrating for 300 ps. Strong adhesion and bonding took place in the contact zone between the two nanocones. As the uniaxial tensile loading was performed, the applied load led to a continuous localized tensile deformation at the contact zone. With increasing tensile deformation, the diameter of the contact zone became thinner until fracturing and separating from one of the nanocones. As noted from the figure, the joint interface was not very clear after surface bonding due to atomic diffusion. It was also found that the adhesion zone with disorderly atoms distribution occurred at the interface of two cones, which was mainly by diffusion and viscous flow bonding. Additionally, one of the nanocones left a number of its atoms adhering to the contacting nanocone after debonding.

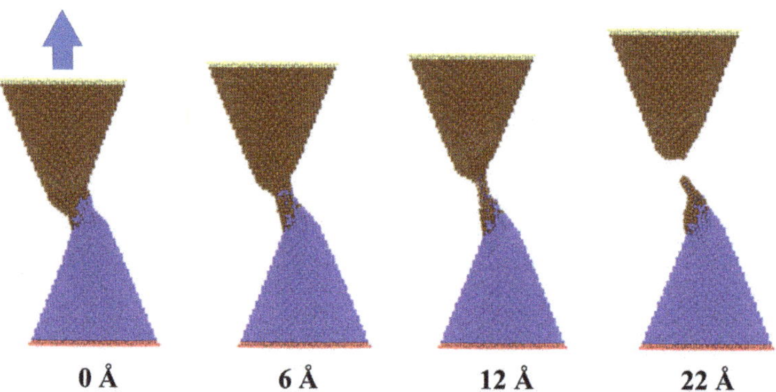

Figure 3. Snapshots of two copper nanocones with an initial axial separation of 20 Å and a contact length of 20 Å at different stages of the uniaxial tensile loading MD simulations.

Figure 4 shows the variation of the axial tensile force exerted on the system and the number of Cu_A-Cu_B bonds via the imposed displacement. As illustrated, the curve can be separated into three regimes. In regime one, the force-displacement behaviors present approximate linearity before yielding, and then the value of tensile force increased gradually with increasing imposed displacement until the curve reaches the peak. In regime two, after the first peak, the tensile force decreased rapidly and then raised to another peak with increasing imposed displacement. Subsequently, the tensile force decreased and reached a local minimum at about 10 Å, then the relationship between tensile force and displacement presented a zigzag pattern. Finally, in regime three, the value of tensile force closed to zero until the two cones separated. It appeared that the interface becomes harder and stronger repeatedly during the necking phase in regime two, which is different from the typical material tensile curve. To further study this phenomenon, the number of Cu_A-Cu_B bonds with a cutoff radius of 3.2 Å was used to describe the formation of cold bonds and surface joints at the interface. From the results, increasing the displacement would affect tensile force while increasing Cu_A-Cu_B bonds in the form of steps. During the bonding process, the top atoms of the nanocones were very active, which not only formed the van der Waals force connection but also indicated the formation of atomic diffusion and cold bonds. In the early stages of stretching, both van der Waals interactions and few created bonds induced elastic deformation. With the increase in deformation, the van der Waals force decreased, and the necking behavior made the interface thinner, which resulted in enhanced surface atom diffusion and intermixing of surface atoms between the interfaces. More cold bonds were formed. It is also noted that more created metal bonds accompanied strain hardening, suggesting that the atomic diffusion is helpful to restrain interface failure.

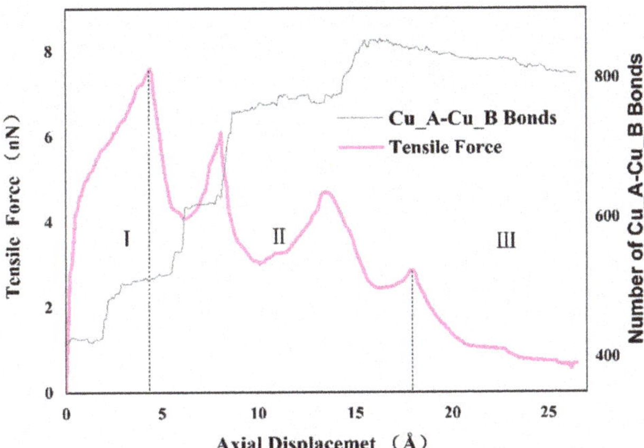

Figure 4. Variation of the tensile loading force and the number of Cu_A-Cu_B bonds with the axial displacement for nanocones, with an initial axial separation of 20 Å and a contact length of 20 Å.

In order to further understand the interface bonding mechanism, the adhesion energy generated by both non-bonded long-range van der Waals force and cold bonds' interaction at the interface between the two cones during debonding process was investigated. In this study, the adhesion energy could be calculated from Equation (1)

$$E_{adhesion} = E_{vdW} + E_{bond} \tag{1}$$

where $E_{adhesion}$ is the total interaction energy between Cu_A and Cu_B, E_{bond} is the Cu_A-Cu_B bonds' interaction energy with a cutoff radius of 3.2 Å, and E_{vdW} is long-range non-bonded van der Waals energy between Cu_A and Cu_B with a cutoff radius of 8.912 Å. Figure 5 shows the adhesion energy and van der Waals energy distributions with different axial displacement during the debonding process. The negative value indicates that the Cu_A and Cu_B atoms attract to each other. The absolute value of the adhesion energy increases with increasing axial displacement. However, the absolute value of van der Waals energy decreases with an increase in axial displacement. The reason is that more cold bonds are formed during the stretching process, as shown in Figure 4, and E_{bond} increases continuously, while the absolute value of long-range van der Waals energy decreases with the increasing distance between Cu_A and Cu_B. It was also found that the contribution of van der Waals energy and metallic bond energy to the adhesion properties is close before the tensile force reaches the maximum. With the further increase in tensile displacement, the bond energy can play a main role in the necking stage.

We also investigated the effect of separation distance and contact length on the joining strength of two copper nanocones with a diameter of 80 Å and a height of 100 Å at a temperature of 300 K. Figure 6 shows the maximum axial tensile loading force that the formed joint between the two nanocones can resist, with the variation of the separation distance. The results show that the maximum axial tensile loading force decreases with the increase in the separation distance. When the separation distance is more than 20 Å, the maximum axial tensile loading force decreases more greatly. This difference should be ascribed to the following reasons: (1) Both van der Waals force and atomic diffusion affect the maximum axial tensile loading force, and the closer the distance is, the greater the influence of atomic diffusion; (2) beyond a certain separation, diffusion ceases and the van der Waals attraction, which at this point plays the dominant role in joint formation, decreases, leading to a steeper decrease in maximum tensile force.

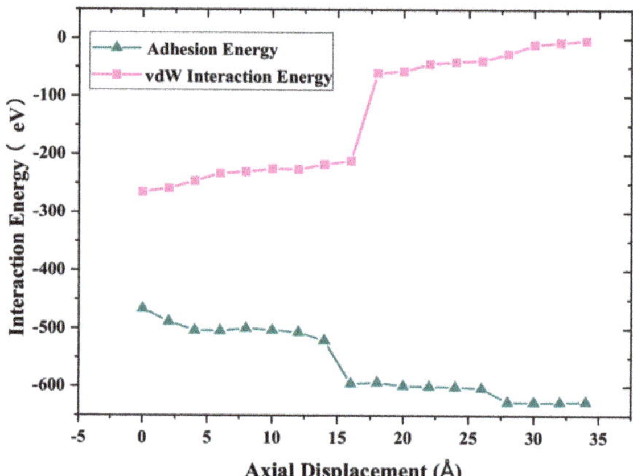

Figure 5. Variation of the adhesion energy and vdW interaction energy with the axial displacement.

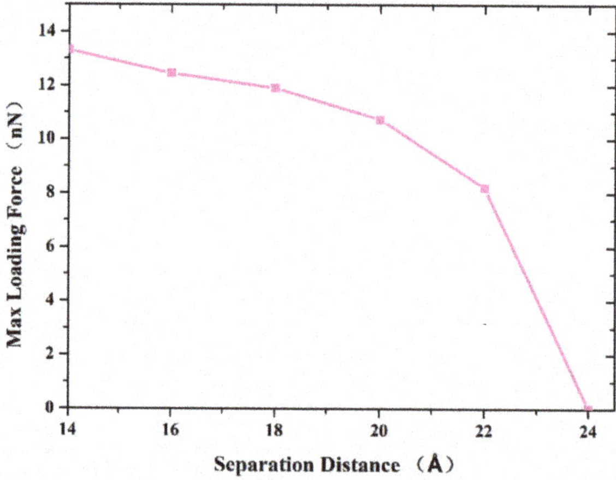

Figure 6. The maximum axial force that the interface can resist with the separation distance.

Figure 7 shows the maximum axial force that the interface can resist with separation distance of 20 Å and various contact length. The curve can be divided into three zones. The first zone, with a contact length less than 15 Å, is concerned with a small interface region and limited adherence of active atoms at the tip of nanocones, as shown in Figure 8a. The slope of the curve in this region is the highest of the three regions. The reason may be ascribed to relatively large numbers of disordered active atom at the cone tips, which diffuse in increasing numbers to form cold bonds as the contact length increases. In the second zone, the increase in maximum axial tensile loading force is due to more atoms on the surface of the cones being involved in forming the joint as shown in Figure 8b, and the increment of the number of metal bonds formed by atomic diffusion is smaller than that of the first zone. Additionally, the increased effective contact area increases the van der Waals force. In the third zone, where the contact length is more than 20 Å, larger contact areas are involved in forming the adherence joint, as shown in Figure 8c. On the other hand, to achieve a larger contact length, larger preloading forces, which will contribute to the diffusion of interface atoms, are typically used. As a result, more cold bonds are formed at the interface so that the slope of the curve in this region is higher than that of the second zone.

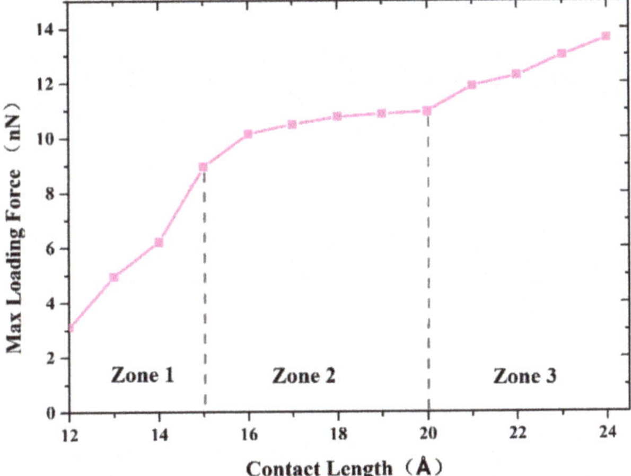

Figure 7. The maximum axial force that the interface can resist with separation distance of 20 Å and various contact length.

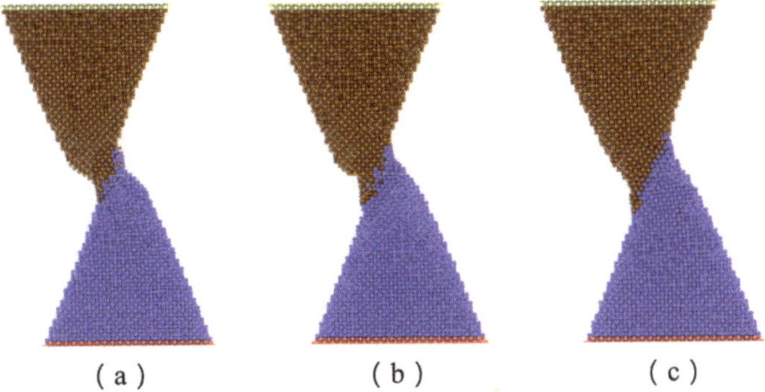

Figure 8. Snapshots of two copper cones forming a bonding interface with a separation distance of 20 Å and various contact lengths: (**a**) 15 Å; (**b**) 20 Å; (**c**) 25 Å.

The effect of temperatures ranging from 300 to 500 K on the strength of the formed joints between two cones was also investigated in this study. Figure 9 shows the variation of the maximum tensile force that the contact nanocones can sustain with increased temperature. For simulation at a constant temperature, the maximum tensile force increases with the increase in temperature as the temperature is lower than 350 °C. Additionally, the maximum tensile force keeps on decreasing with the increase in temperature. It appeared that the increase in temperature could significantly promote the diffusion of interface atoms, resulting in increasing joint strength. However, when the temperature was increased, the maximum tensile forces were continuously decreased. This suggested that higher temperature promotes melting of the joint, and therefore weak resistance to a tensile force. On the other hand, when the adhesive joint was cooled to 300 °C before debonding, the maximum tensile force kept on increasing with the increase in temperature, as shown in Figure 7. This result illustrates the significant effect of temperature on the performance of surface bonding based on metal nanocones, and will thus help to tune the joint strength for high application.

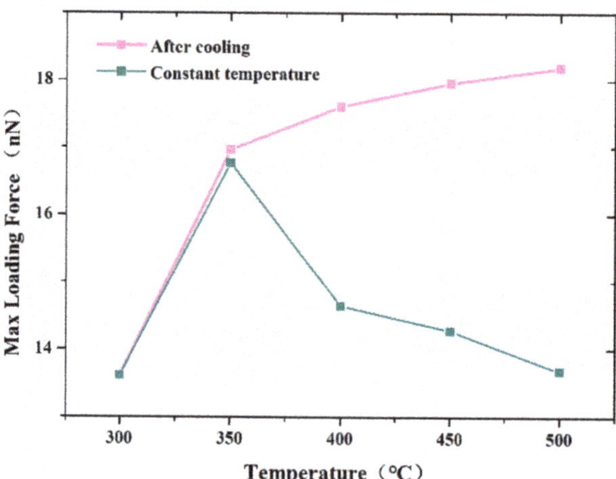

Figure 9. Effect of temperature on the strength of the formed joints between two cones.

As mentioned above, both atom diffusion and Van der Waals force depending on contact configuration determine the joint strength. Therefore, various copper nanocones, coupled with different diameters and heights, were investigated to find out the effect of surface structure on bonding strength. Serial simulations were performed with a diameter of 80 Å, a separation distance of 20 Å, and a contact length of 20 Å. Figure 10 shows the variation in the maximum tensile force with different heights. The maximum tensile strength increases when the height is lower than 80 Å. However, the maximum tensile strength reaches its highest value at a cone height of about 80 Å, which corresponds to a cone angle of 53°, and decreases thereafter. From a geometric point of view, a larger effective contact area and shorter vertical distance between two surfaces were obtained, with an increasing cone angle corresponding to a lower height diameter ratio. This helps to improve the interface bonding strength. On the other hand, a larger cone angle means that there are relatively few disordered active atoms on the tip, which will reduce atoms' diffusion during the formation of the joint.

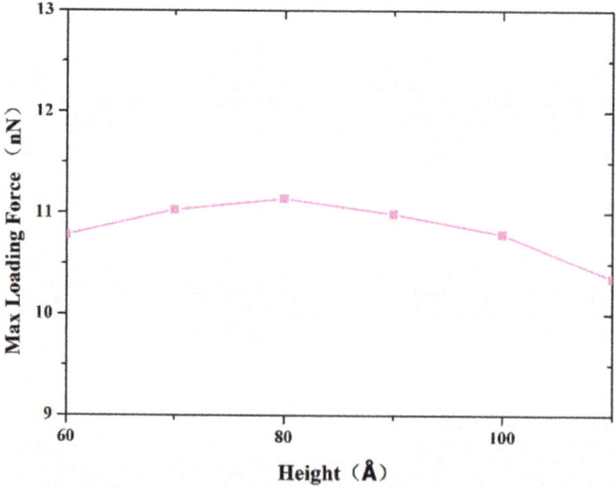

Figure 10. Effect of height of the cone on the strength of the formed joints between two cones with diameter 80 Å, separation distance 20 Å, and contact length 20 Å.

4. Conclusions

In this study, we developed a molecular dynamics analysis model of surface bonding based on metal nanocones arrays. The influences of separation distance, contact length, temperature, and size of the cones were examined. Specifically, the focus of this study was devoted to examining and characterizing the resistance capacity to tension force and failure mechanisms associated with the above parameters. The results reveal that both van der Waals interactions and surface atoms' diffusion are essential constituents that determine the bonding strength. The debonding process of different failure mechanisms is related to their deformation behavior in different regimes. Noteworthy is the fact that more created metal bonds accompanied strain hardening during debonding, which is helpful to restrain interface failure. It is also shown that increasing contact length and decreasing separation distance can simultaneously contribute to increasing the bonding strength. Additionally, increasing the operating temperature and subsequent cooling would contribute to atoms' diffusion and bonding strength increasing. It also reveals that the maximum bonding strength was obtained when the cone angle was around 53°. Decidedly, the work offers bonding and debonding mechanisms involved at the atomic level for surface bonding based on metal nanocone arrays, which is helpful to design and optimize the surface bonding process.

Author Contributions: Conceptualization, X.S. and R.Z.; methodology, X.S. and R.Z.; software, X.S. and S.W.; validation, S.W.; formal analysis, X.S.; investigation, X.S. and R.Z.; resources, R.Z.; data curation, S.W.; writing—original draft preparation, X.S.; writing—review and editing, R.Z.; visualization, S.W.; supervision, R.Z.; project administration, R.Z.; funding acquisition, X.S. All authors have read and agreed to the published version of the manuscript.

Funding: This study was funded by Zhongyuan Science and Technology Innovation Leadership Program of China (Grant No. 214200510014).

Data Availability Statement: Data available in a publicly accessible repository.

Conflicts of Interest: The authors declare no conflict of interest.

References

1. Shin, H.; Han, E.; Park, N.; Kim, D. Thermal residual stress analysis of soldering and lamination processes for fabrication of crystalline silicon photovoltaic modules. *Energies* **2018**, *11*, 3256. [CrossRef]
2. Vianco, P.T. A review of interface microstructures in electronic packaging applications: Soldering technology. *JOM* **2019**, *71*, 158–177. [CrossRef]
3. Ko, C.T.; Chen, K.N. Low temperature bonding technology for 3d integration. *Microelectron. Reliab.* **2012**, *52*, 302–311. [CrossRef]
4. Ju, Y.; Amano, M.; Chen, M. Mechanical and electrical cold bonding based on metallic nanowire surface fasteners. *Nanotechnology* **2012**, *23*, 365202. [CrossRef] [PubMed]
5. Wang, P.; Ju, Y.; Chen, M.; Hosoi, A.; Song, Y.; Iwasaki, Y. Room-temperature bonding technique based on copper nanowire surface fastener. *Appl. Phys. Express* **2013**, *6*, 035001. [CrossRef]
6. Li, J.; Yu, X.; Shi, T.; Cheng, C.; Fan, J.; Cheng, S.; Liao, G.; Tang, Z. Low-temperature and low-pressure cu–cu bonding by highly sinterable cu nanoparticle paste. *Nanoscale Res. Lett.* **2017**, *12*, 1–6. [CrossRef]
7. Wang, H.; Leong, W.S.; Hu, F.; Ju, L.; Su, C.; Guo, Y.; Li, J.; Li, M.; Hu, A.; Kong, J. Low-temperature copper bonding strategy with graphene interlayer. *ACS Nano* **2018**, *12*, 2395–2402. [CrossRef]
8. Lee, J.; Lee, I.; Kim, T.S.; Lee, J.Y. Efficient welding of silver nanowire networks without post-processing. *Small* **2013**, *9*, 2887–2894. [CrossRef] [PubMed]
9. Toku, Y.; Ichioka, K.; Morita, Y.; Ju, Y. A 64-pin nanowire surface fastener like a ball grid array applied for room-temperature electrical bonding. *Sci. Rep.* **2019**, *9*, 1–10. [CrossRef] [PubMed]
10. Wang, P.; Yang, J. Room-temperature electrical bonding technique based on copper/polystyrene core/shell nanowire surface fastener. *Appl. Surf. Sci.* **2015**, *349*, 774–779. [CrossRef]
11. Yuhki, T.; Keita, U.; Yasuyuki, M.; Yang, J. Nanowire surface fastener fabrication on flexible substrate. *Nanotechnology* **2018**, *29*, 305702.
12. Wang, P.; Ju, Y.; Cui, Y.; Hosoi, A. Copper/parylene core/shell nanowire surface fastener used for room-temperature electrical bonding. *Langmuir* **2013**, *29*, 13909–13916. [CrossRef]
13. Kang, Z.; Wu, B. Coalescence of gold nanoparticles around the end of a carbon nanotube: A molecular-dynamics study. *J. Manuf. Process.* **2018**, *34*, 785–792. [CrossRef]

14. Cha, S.H.; Park, Y.; Han, J.W.; Kim, K.; Kim, H.S.; Jang, H.L.; Cho, S. Cold welding of gold nanoparticles on mica substrate: Self-adjustment and enhanced diffusion. *Sci. Rep.* **2016**, *6*, 32951. [CrossRef]

15. Mo, Y.; Turner, K.T.; Szlufarska, I. Friction laws at the nanoscale. *Nature* **2009**, *457*, 1116–1119. [CrossRef] [PubMed]

16. Xu, P.; Hu, F.; Shang, J.; Hu, A.; Li, M. An ambient temperature ultrasonic bonding technology based on cu micro-cone arrays for 3d packaging. *Mater. Lett.* **2016**, *176*, 155–158. [CrossRef]

17. Wang, H.; Ju, L.; Guo, Y.; Mo, X.; Ma, S.; Hu, A.; Li, M. Interfacial morphology evolution of a novel room-temperature ultrasonic bonding method based on nanocone arrays. *Appl. Surf. Sci.* **2015**, *324*, 849–853. [CrossRef]

18. Li, Q.; Fu, T.; Peng, T.; Peng, X.; Liu, C.; Shi, X. Coalescence of cu contacted nanoparticles with different heating rates: A molecular dynamics study. *Int. J. Mod. Phys. B Condens. Matter Phys. Stat. Phys. Appl. Phys.* **2016**, *30*, 1650212. [CrossRef]

19. Plimpton, S. Fast parallel algorithms for short-range molecular dynamics. *J. Comput. Phys.* **1995**, *117*, 1–19. [CrossRef]

20. Stukowski, A. Visualization and analysis of atomistic simulation data with OVITO–the Open Visualization Tool. *Model. Simul. Mater. Sci. Eng.* **2010**, *18*, 015012. [CrossRef]

21. Ren, X.; Li, X.; Huang, C.; Yin, H.; Wei, F. Molecular dynamics simulation of thermal welding morphology of ag/au/cu nanoparticles distributed on si substrates. *Ferroelectrics* **2020**, *564*, 19–27. [CrossRef]

22. Heinz, H.; Vaia, R.A.; Farmer, B.L.; Naik, R.R. Accurate simulation of surfaces and interfaces of face-centered cubic metals using 12-6 and 9-6 lennard-jones potentials. *J. Phys. Chem. C* **2008**, *112*, 17281–17290. [CrossRef]

23. Foiles, S.M.; Baskes, M.I.; Daw, M.S. Embedded-atom-method functions for the fcc metals Cu, Ag, Au, Ni, Pd, Pt, and their alloys. *Phys. Rev. B* **1986**, *33*, 7983–7991. [CrossRef]

24. Mishin, Y.; Mehl, M.; Papaconstantopoulos, D.; Voter, A.; Kress, J. Structural stability and lattice defects in copper: Ab initio, tight-binding, and embedded-atom calculations. *Phys. Rev. B* **2001**, *63*, 224106. [CrossRef]

25. Hao, H.; Lau, D. Atomistic modeling of metallic thin films by modified embedded atom method. *Appl. Surf. Sci.* **2017**, *422*, 1139–1146. [CrossRef]

26. Weng, S.; Ning, H.; Fu, T.; Hu, N.; Zhao, Y.; Huang, C.; Peng, X. Molecular dynamics study of strengthening mechanism of nanolaminated graphene/cu composites under compression. *Sci. Rep.* **2018**, *8*, 1–10. [CrossRef] [PubMed]

27. Hansson, P. Influence of surface roughening on indentation behavior of thin copper coatings using a molecular dynamics approach. *Comput. Mater. Sci.* **2016**, *117*, 233–239. [CrossRef]

28. Zhou, H.; Li, J.; Xian, Y.; Hu, G.; Li, X.; Xia, R. Nanoscale assembly of copper bearing-sleeve via cold-welding: A molecular dynamics study. *Nanomaterials* **2018**, *8*, 785. [CrossRef] [PubMed]

29. Liu, T.; Liu, G.; Wriggers, P.; Zhu, S. Study on contact characteristic of nanoscale asperities by using molecular dynamics simulations. *J. Tribol.* **2009**, *131*, 1–10. [CrossRef]

30. Jung, S.C.; Suh, D.; Yoon, W.S. Molecular dynamics simulation on the energy exchanges and adhesion probability of a nano-sized particle colliding with a weakly attractive static surface. *J. Aerosol Sci.* **2010**, *41*, 745–759. [CrossRef]

31. Pierre, H. A tool for manipulating and converting atomic data files. *Comput. Phys. Comm.* **2015**, *197*, 212–219.

32. Hickman, J.; Mishin, Y. Temperature fluctuations in canonical systems: Insights from molecular dynamics simulations. *Phys. Rev. B* **2016**, *94*, 184311. [CrossRef]

 *nanomaterials*

MDPI

Article

Molecular Dynamics Simulation for the Effect of Fluorinated Graphene Oxide Layer Spacing on the Thermal and Mechanical Properties of Fluorinated Epoxy Resin

Qijun Duan [1,2], Jun Xie [1,*], Guowei Xia [1], Chaoxuan Xiao [1], Xinyu Yang [1], Qing Xie [1,2] and Zhengyong Huang [3]

[1] Hebei Provincial Key Laboratory of Power Transmission Equipment Security Defense, North China Electric Power University, Baoding 071003, China; duan_ncepu@163.com (Q.D.); xgw_97@163.com (G.X.); XCX@ncepu.edu.cn (C.X.); yang_HD@126.com (X.Y.); xq_ncepu@126.com (Q.X.)
[2] State Key Laboratory of Alternate Electrical Power System with Renewable Energy Sources, North China Electric Power University, Beijing 102206, China
[3] State Key Laboratory of Power Transmission Equipment & System Security and New Technology, Chongqing University, Chongqing 400044, China; huangzhengyong@cqu.edu.cn
* Correspondence: junxie@ncepu.edu.cn

Citation: Duan, Q.; Xie, J.; Xia, G.; Xiao, C.; Yang, X.; Xie, Q.; Huang, Z. Molecular Dynamics Simulation for the Effect of Fluorinated Graphene Oxide Layer Spacing on the Thermal and Mechanical Properties of Fluorinated Epoxy Resin. *Nanomaterials* **2021**, *11*, 1344. https://doi.org/10.3390/nano11051344

Academic Editors: Francesco Tornabene and Rossana Dimitri

Received: 23 April 2021
Accepted: 17 May 2021
Published: 20 May 2021

Abstract: Traditional epoxy resin (EP) materials have difficulty to meet the performance requirements in the increasingly complex operating environment of the electrical and electronic industry. Therefore, it is necessary to study the design and development of new epoxy composites. At present, fluorinated epoxy resin (F-EP) is widely used, but its thermal and mechanical properties cannot meet the demand. In this paper, fluorinated epoxy resin was modified by ordered filling of fluorinated graphene oxide (FGO). The effect of FGO interlayer spacing on the thermal and mechanical properties of the composite was studied by molecular dynamics (MD) simulation. It is found that FGO with ordered filling can significantly improve the thermal and mechanical properties of F-EP, and the modification effect is better than that of FGO with disordered filling. When the interlayer spacing of FGO is about 9 Å, the elastic modulus, glass transition temperature, thermal expansion coefficient, and thermal conductivity of FGO are improved with best effect. Furthermore, we calculated the micro parameters of different systems, and analyzed the influencing mechanism of ordered filling and FGO layer spacing on the properties of F-EP. It is considered that FGO can bind the F-EP molecules on both sides of the nanosheets, reducing the movement ability of the molecular segments of the materials, so as to achieve the enhancement effect. The results can provide new ideas for the development of high-performance epoxy nanocomposites.

Keywords: fluorinated epoxy resin; fluorinated graphene oxide; ordered filling; molecular dynamics; elastic modulus; glass transition temperature; microscopic parameters

1. Introduction

Epoxy resin reacts with a curing agent to form a polymer with a three-dimensional network structure. The cured product has excellent electrical insulation, mechanical properties, and chemical corrosion resistance. Therefore, EP materials are widely used in electrical insulation, electronics, aerospace, machinery, and construction [1–4]. As the operating conditions of epoxy resin materials become increasingly complex, higher requirements have been put forward for their thermal, mechanical, and insulation properties [5,6]. In recent years, fluorinated epoxy resin (F-EP) has gradually attracted researchers' attention due to excellent insulation and dielectric properties coming from the extremely strong electronegativity of fluorine and the high bond energy of C-F bond [7–9]. However, the mechanical strength, heat resistance, and thermal conductivity of F-EP materials cannot meet the requirements of high voltage insulation materials, which has become a key factor restricting its further application. Therefore, it is necessary to improve its thermal and mechanical properties.

Nano modification is one of the significant means to improve the properties of epoxy composites. With the continuous advancement of the research and application of carbon nanomaterials, modified graphene materials doped with epoxy resin have become a hot spot in current research [10–13]. Graphene oxide is a graphene-based material containing a large number of oxidizing functional groups obtained after graphite is oxidized. It retains most of the excellent properties of graphene and has high surface activity. However, due to the strong van der Waals force [14] on the surface of graphene-based materials, graphene oxide flakes in the composites are easy to form serious aggregates. Relevant experimental studies also confirmed that the distribution of graphene directly affects the macro properties of the composites [15–17]. It is found that the dispersion of nano fillers in polymer matrix can be improved by fluorination modification, and the fluorine-containing groups will form a shielding layer on the surface of nano materials, so as to inhibit the agglomeration of fillers [18–20]. This provides a method guidance for the construction of a modified graphene packing network with good dispersion.

One dimensional or two-dimensional nano fillers are often affected by their own spatial structure in many aspects. However, the agglomeration and disordered distribution of nano fillers will seriously affect the modification effect [21–23]. The researchers found that reasonable assembly of fillers in the polymer matrix has better modification effect on the properties of composites [24–26]. In addition, the physical and chemical properties of nano fillers have an important influence on the cross-linking structure and crystallization behavior of polymers, but the mechanism has not been revealed, and the research on this problem is rare. Sanat K. Kumar et al. reviewed the outstanding theoretical research progress of polymer-nanoparticle hybrids, focusing on the functionalization of nanomaterials and self-assembly methods of fillers, and pointed out that the design of the nanofiller network is a key point of future research [27]. Professor Ahmad Jabbarzadeh has conducted in-depth and systematic research on the crystallization behavior of nanocomposite polymers. The effects of the shape, size, and volume fraction of fillers on the crystallization behavior of the polymer were analyzed, and the crystallization mechanism of the nanocomposite polymer was revealed [28,29]. These studies provide strong support for the development of high-performance functional nanocomposites. The two-dimensional properties of modified graphene make it easy to construct the filler network with layered structure [30,31]. However, there is still a gap in the research on the ordered filling of functional graphene nanosheets in epoxy resin materials. Moreover, it is difficult to fix the space position of modified graphene by experimental method, so molecular simulation technology can be used to realize the pre-study of this method.

At present, molecular dynamics (MD) simulation technology has been widely used in the design, development, and performance analysis of materials [32–35]. The MD technology can be used to simulate the structure and properties of high molecular polymers from the molecular scale, and analyze the correlation between the microstructure and macroscopic properties of the polymer. MD technology has been widely used by researchers in the development and research of epoxy composite materials, which not only saves research time and economic costs, but also deepens the understanding of the modification mechanism of epoxy composite materials [36–38]. Shenogina [39] built a diglycidyl ethers bisphenol A (DGEBA)/diethyltoluenediamine (DETDA) cross-linking network system through MD simulation technology, and studied the influence of the number of molecules, molecular chain length, and cross-linking degree on the thermal performance of the cross-linking network. Wang [40] et al. used MD simulation to modify graphene with functional groups, and explored the effect of graphene modification methods on the Young's modulus and thermal conductivity of composites. The results showed that the carboxyl and amine functionalized graphene nanocomposites have optimal performance, and the numerical results are in good agreement with the experimental data.

In this study, we established a model of fluorinated graphene oxide (FGO)/F-EP composites with layer spacing of 3 Å, 6 Å, 9 Å, 12 Å, and disorderly filling. Based on molecular dynamics simulation technology, the effects of different distribution characteristics of FGO nanosheets on the thermal and mechanical properties of the composites were studied. This study provides a new theoretical understanding for epoxy resin reinforced by modified graphene, which can guide the design and development of high-performance epoxy nanocomposites.

2. Model Construction and Simulation Calculation

In this study, we used Material Studio 7.0 to complete the modeling and calculation. Firstly, the cross-linking network model of fluorine-containing epoxy resin was constructed, and the DGEBA was fluorinated with hexafluorophobia bisphenol A (BPAF) as a fluorinating agent [8,41]. The monomer molecular models of DGEBA, methyltetrahydrophthalic anhydride (MTHPA), and BPAF were constructed. Each monomer molecule was labeled with reactive atoms and optimized in MD geometry, as shown in Figure 1a. The optimized monomer molecules were put into an amorphous periodic box according to the ratio in Table 1. The box density was set at 0.6 g/cm^3, the model temperature was 600 K, and the model was dynamically optimized. Furthermore, graphene unit cells were introduced, and graphene supercells were constructed according to the periodic box size. The molecular model of graphene oxide (GO) monomer was constructed based on Lerf–Klinowski method, and the simplest molecular formula was $C_{10}O_1(OH)_1(COOH)_{0.5}$. In the model, epoxy groups and hydroxyl groups are randomly attached to the surface, while carboxyl groups are distributed at the edges. In this paper, a GO sheet containing 82 carbon atoms and 22 oxygen atoms is constructed to represent graphene oxide, and hydrogen atoms are added to prevent unsaturated edges. The GO model is shown in Figure 1a. Then, the graphene oxide is fluorinated, and fluorine atoms are manually added at its edges to construct a FGO model.

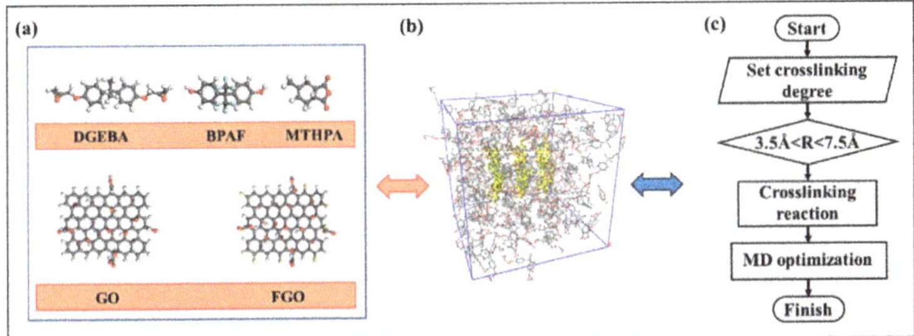

Figure 1. Establishment of molecular models. (**a**) Monomer molecular models of DGEBA, BPAF, MTHPA, GO, and FGO. (**b**) Model of F-EP filled with FGO. (**c**) Crosslinking procedure.

Table 1. Composition of FGO modified F-EP Composites.

System	Layer Spacing (Å)	Number of Molecules			
		FGO	DGEBA	MTHPA	BPAF
F-EP	–	0	50	100	25
Random	–	3	50	100	25
3-FGO/F-EP	3	3	50	100	25
6-FGO/F-EP	6	3	50	100	25
9-FGO/F-EP	9	3	50	100	25
12-FGO/F-EP	12	3	50	100	25

The FGO nanosheets were filled into the F-EP box to construct the composite material model, as shown in Figure 1b. The molecular dynamics calculation results of epoxy resin are greatly affected by the number of model molecules, so we control the same number of molecules in each model, and only change the interlayer spacing of modified graphene nanosheets. We filled three FGO Nanosheets in the cell and controlled the FGO Nanosheets array to be located in the center of the cell. The distance between FGO layers was set as 3 Å, 6 Å, 9 Å, and 12 Å, respectively. The position coordinates of FGO molecules in the model were fixed, and a disordered filling model was constructed as the control group. The unfilled F-EP model and the FGO filled composites models with hidden epoxy resin are shown in Figure 2.

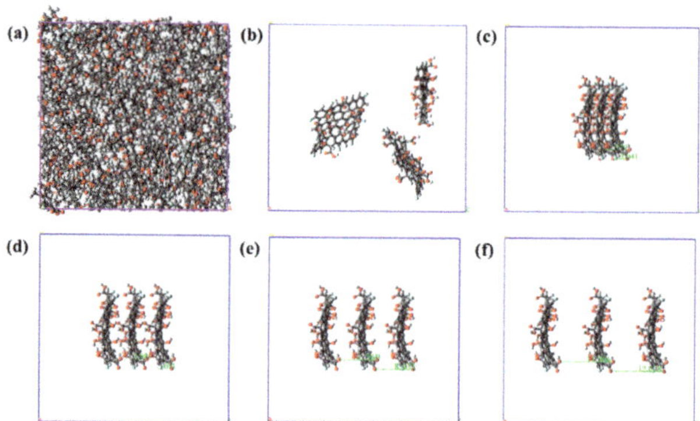

Figure 2. Molecular models. (**a**) Unfilled F-EP. (**b**) F-EP with disorderly filled FGO nanosheets. Figure (**c**) to (**e**) are the F-EP composite models with ordered filling of FGO (epoxy resin molecules are hidden). The layer spacing of each model is as follows: (**c**) 3 Å, (**d**) 6 Å, (**e**) 9 Å, (**f**) 12 Å.

The geometric optimization of the composite model with different FGO layer spacing was carried out. The box with the lowest energy was selected to calculate the binding energy parameters of GO, FGO, and F-EP. After the nano-filler is doped into the epoxy matrix, an inorganic−organic interface layer will be formed between filler and matrix. Generally, the stronger the interfacial force between the nano-filler and the matrix, the better the performance of the corresponding composite. Binding energy is an important parameter to characterize the interfacial bonding force between modified filler and epoxy system. The larger its absolute value, the stronger the interaction force between matrix and filler in composite material [42], and the calculation formula is as follows:

$$E_{interface} = E_{total} - E_{fiber} - E_{resin} \tag{1}$$

where $E_{interface}$ is the interfacial bonding energy between nanofiller and matrix, E_{total} is the total energy of composite material, E_{resin} is the energy of epoxy substrate, and E_{fiber} is the energy of nano-filler. The calculation results of bonding energy are shown in Table 2.

Table 2. Binding energy of different modified graphene/F-EP systems (kcal/mol).

System	E_{resin}	E_{fiber}	E_{total}	$E_{interface}$
GO/F-EP	−4761	−769	−4157	1373
FGO/F-EP	−4761	−805	−3766	1799

As illustrated in Table 2, the interfacial bonding energy between FGO and matrix after grafting fluorine element is obviously improved compared with GO. The analysis shows that fluorine has strong electronegativity, and bonding with carbon will make the common electrons of fluorocarbon atoms tend to fluorine atoms, forming a negative charge shielding layer, which inhibits the agglomeration effect of modified graphene materials and provides more surface area for the interaction between filler and substrate. In addition, due to the strong polarity of C-F bond, it is easy to react with groups in epoxy resin, and the interfacial bonding strength between filler and epoxy matrix can be enhanced at low filling mass fraction. Therefore, the fluorinated grafting modification of graphene oxide nanosheets is beneficial to improve its dispersibility and compatibility in epoxy resin matrix.

The model of composites with cross-linking degree of 90% was obtained by further running the cross-linking program [4], the calculation program is shown in Figure 1c. The cross-linking temperature was set at 600 K, and the truncation radius of reaction atom bonding was set at 3.5 Å−7.5 Å. After cross-linking, the model was optimized geometrically and MD treated to eliminate the internal stress generated during cross-linking. In MD process, NVT of 100 ps and NPT relaxation of 200 ps were carried out at 600 K, the Andersen and Berendsen were chosen to control temperature and pressure, respectively. The pressure of molecular dynamics process is 10^5 Pa and the time step is 1 fs. The size of the cross-linking network model of the optimized epoxy resin composite is 43 Å*43 Å*43 Å. The optimized composite models were annealed, with the annealing span controlled at 600–280 K and the annealing rate selected as 20 K/100 ps. After each round of annealing, NPT optimization treatment of 200 ps was performed to eliminate the internal stress caused by temperature change. Finally, the epoxy composite models at different temperatures were output for the subsequent calculation of system performance and structural parameters.

3. Results and Discussion

3.1. Static Mechanical Performance Calculation

In this paper, the static constant strain method was used to analyze the elastic mechanical properties of the system [43]. After MD optimization process, the system had reached the mechanical equilibrium. Then, a small strain was applied to it and the energy optimization was carried out again. In the process of molecular simulation, the strain is applied to different directions and repeated many times. The stiffness matrix can be calculated as the second derivative of the deformation energy (U) per unit volume (V) with respect to strain (ε).

$$C_{ij} = \frac{1}{V} * \frac{\partial^2 U}{\partial \varepsilon_i \partial \varepsilon_j} \tag{2}$$

In this process, three groups of uniaxial tension, three groups of uniaxial compression, and six groups of shear deformation were applied to the balanced epoxy resin system in x, y, and z directions respectively. The stiffness constant matrix can be obtained by extracting the stress of the optimized deformation system. We found that the elastic mechanical properties of the matrix did not show obvious anisotropy, which is related to the small size and mass fraction of FGO. Therefore, we calculated elastic mechanical parameters according to the isotropic stiffness matrix of conventional epoxy composites.

$$C_{ij} = \begin{bmatrix} \lambda+2\mu & \lambda & \lambda & 0 & 0 & 0 \\ \lambda & \lambda+2\mu & \lambda & 0 & 0 & 0 \\ \lambda & \lambda & \lambda+2\mu & 0 & 0 & 0 \\ 0 & 0 & 0 & \mu & 0 & 0 \\ 0 & 0 & 0 & 0 & \mu & 0 \\ 0 & 0 & 0 & 0 & 0 & \mu \end{bmatrix} \tag{3}$$

where λ and μ are elastic constants, and the corresponding constants can be obtained by further solving the matrix.

$$\begin{cases} \lambda = \frac{1}{6}(C_{12} + C_{13} + C_{21} + C_{23} + C_{31} + C_{32}) \\ \mu = \frac{1}{3}(C_{44} + C_{55} + C_{66}) \end{cases} \tag{4}$$

According to λ and μ, the static elastic modulus of the system such as Young's modulus E, shear modulus G, and bulk modulus K can be obtained.

$$\begin{cases} E = \mu\frac{3\lambda+2\mu}{\lambda+\mu} \\ G = \mu \\ K = \lambda + \frac{2}{3}\mu \end{cases} \tag{5}$$

In this study, the static elastic modulus of each model at 300 K is calculated, and the results are shown in Figure 3. It can be found that when the interlayer spacing is changed from 3 Å to 12 Å, the mechanical properties of the material show a sinusoidal-like behavior of first decreasing and then increasing. When the spacing between FGO layers is 9 Å, the static elastic modulus of the composite reaches the highest, and it can be considered that the comprehensive mechanical properties of the composite are better at this time. It is considered that when the interlayer spacing is small, the structure distribution of the composite material is uneven, which cannot utilize the characteristics of high elastic modulus of modified graphene. However, when the interlayer spacing is too large, the van der Waals interaction force between FGO and epoxy matrix is weak, which leads to the weak deformation resistance of the system. After adjusting the distance between the layers, the modified graphene can fully contact with the matrix material and form a well dispersed adsorption layer. The external stress on the substrate can be effectively buffered in the adsorption layer and transferred to the modified graphene sheet with excellent mechanical properties. At the same time, the force of FGO on the matrix also limits the movement of a polymer molecular chain, which makes the epoxy resin form a relatively close cross-linking network. This can also better bear the stress of the material. Therefore, reasonable control of FGO filler network structure can improve the mechanical properties of the composites.

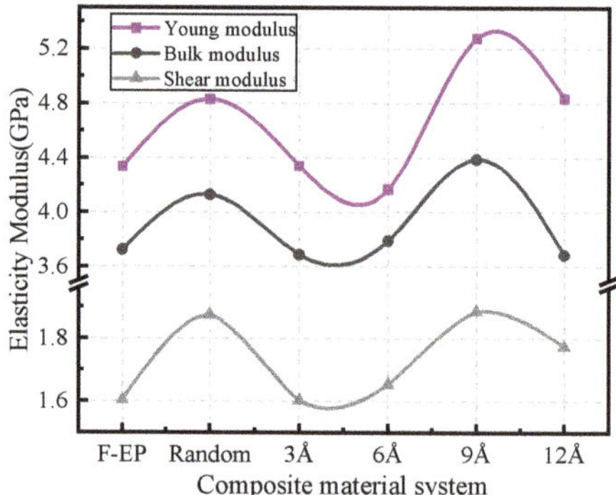

Figure 3. Static elastic modulus of FGO/F-EP composites with different distribution characteristics.

3.2. Thermal Performance Analysis

3.2.1. Glass Transition Temperature

Glass transition temperature (T_g) can be calculated by extracting the density and temperature parameters of epoxy composites during annealing. During the transition from glassy state to the rubbery state of polymer, the density and volume of materials will change with the increase of temperature. There are obvious differences between the change rates of density and volume of materials with temperatures before and after the glass phase transition [44]. According to this rule, we can get the T_g of epoxy composites by fitting the temperature-density curve, and the change of T_g of FGO/F-EP with different interlayer spacing, as shown in the Figure 4. It can be found that, similar to the change of static elastic modulus, T_g also shows a sinusoidal change with the increase of the interlayer distance. The T_g value reaches the highest when the interlayer distance is 9 Å, which is 483 K. We believe that when the interlayer spacing is small, FGO exists in epoxy substrate in a form similar to agglomerated nano-filler, which has an adverse effect on the improvement of T_g of epoxy composites. With the increase of spacing between FGO layers, FGO nanosheets with fixed spatial positions form a localized filler network with pivotal effect, and F-EP is attracted and fixed near the filler to a certain extent, forming a relatively stable cross-linked network structure. Meanwhile, fluorine atoms existing in both filler and matrix provide a more stable acting force for this connection form, which delays the glass phase transition process of the composites and improves T_g. Similarly, when the interlayer spacing is too large, the pivotal effect of the filler is weakened, and the advantages of the filler itself and the network structure regulation cannot be fully exploited, which makes the T_g a downward trend.

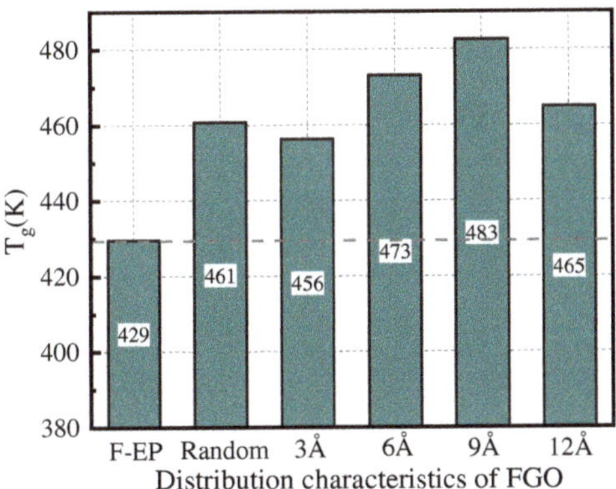

Figure 4. Glass transition temperature of FGO/F-EP composites with different distribution characteristics.

3.2.2. Coefficient of Thermal Expansion

Coefficient of thermal expansion (CTE) is an important parameter to characterize the stability of polymer materials at high temperature, which can be calculated by calculating the internal stress generated after the internal deformation of the system [45]. The calculation formula is:

$$\text{CTE} = \frac{1}{V_0}\left(\frac{\partial V}{\partial T}\right)_P \tag{6}$$

where V_0 is the volume of the initial model after cross-linking (in this study, the volume parameter of the system at 300 K is used), and P is the standard atmospheric pressure. According to the formula (5), $\left(\frac{\partial V}{\partial T}\right)_P$ can be obtained by linear fitting the volume and temperature parameters of the model. Furthermore, the CTE of FGO/F-EP epoxy composites with different interlayer spacing can be obtained, and the results are shown in the Figure 5. It can be found that the overall CTE of the composite material system after adding FGO is lower than that of the pure resin system. However, when the distance between the FGO layers is near the range of 3 Å~6 Å, the filler spacing is too small, resulting in agglomeration. At this time, the filler lacks binding effect on the epoxy resin. As the temperature increases, the heated movement of the filler makes agglomerates expanded, which leads to a slight increase in the thermal expansion coefficient of the composite material. When the interlayer spacing of FGO nanosheets is about 9 Å, the two-dimensional nanosheets have a greater binding effect on the polymer segments and are not affected by filler agglomeration. When the composite material is thermally expanded, the segments preferentially fill the free volume space, so the thermal expansion coefficient of the system is small.

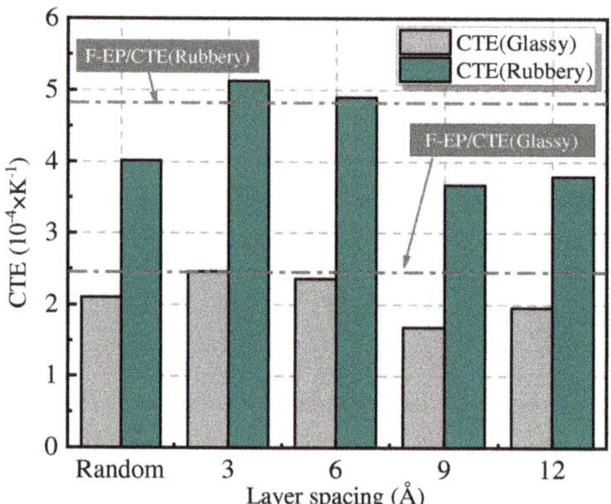

Figure 5. Coefficient of thermal expansion of FGO/F-EP composites with different distribution characteristics.

3.2.3. Thermal Conductivity

In order to study the influence of FGO interlayer distance on the axial thermal conductivity (TC) of materials, the TC of different systems was studied by using TC script. In this script, the TC is calculated according to reverse perturbation nonequilibrium molecular dynamics (RNEMD), and its computer model is shown in Figure 6 [46]. The calculation formula of thermal conductivity is:

$$\kappa = -\frac{\sum\limits_{\text{transfers}} \frac{m}{2}\left(v_h^2 - v_c^2\right)}{2tL_xL_y\langle\partial T/\partial z\rangle} \tag{7}$$

where κ is the exchange rate of hot particles, v_c is the exchange rate of cold particles, L_xL_y is the area where heat transfer occurs, and $\frac{\partial T}{\partial Z}$ is the temperature gradient in the z direction.

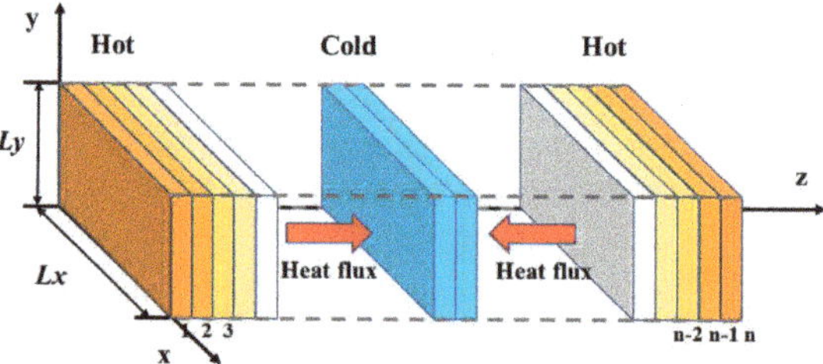

Figure 6. Mechanism model for heat conduction calculation of RNEMD.

In order to improve the accuracy of the simulation calculation, the composite material model needs to be extended three times in the Z-axis direction before calculating the TC to obtain a $1 \times 1 \times 3$ composite unit cell [47]. The TC of each system is calculated 10 times and the average value is taken. The results are shown in the Figure 7 below.

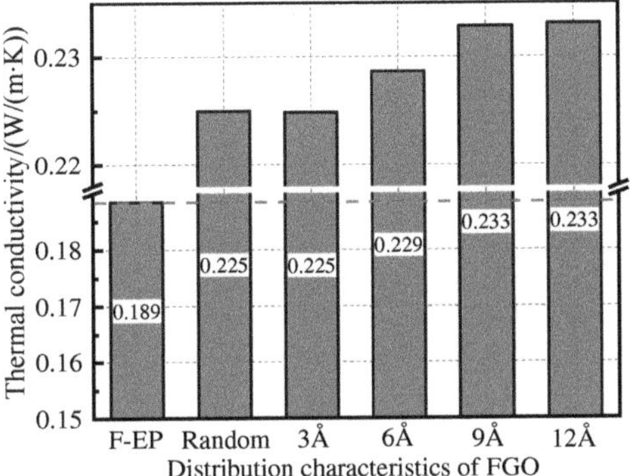

Figure 7. Thermal conductivity of FGO/F-EP composites with different distribution characteristics.

Comparing the TC data of different distribution states of FGO/F-EP system, it can be seen that the orderly-filled FGO can greatly improve the TC of the F-EP system. When the distance between the modified graphene layers is greater than 9 Å, the TC of the system is increased by about 4% relative to the disordered filler distribution system. This shows that graphene itself has good TC, but the commonly used disorderly doping methods often make fillers agglomerate and stack together, which reduces the interaction area between fillers and matrix. At the same time, there is a lack of binding force for the polymer segments outside the filler interface, which makes it difficult to overcome the interfacial thermal resistance between the filler and the matrix in the heat transfer process. As a result, even if the nano filler is doped with high thermal conductivity, the improvement of the thermal conductivity of the composite is very limited.

After FGO is filled in an orderly manner, the uniformly dispersed two-dimensional FGO nanosheets provide a sufficient surface area of the filler, and the presence of fluorine also greatly enhances the bonding between the filler and the matrix. In addition, the polymer segments are bound by spatially fixed fillers, which makes the cross-linking network structure of epoxy resin molecules between nanosheets more compact. These factors play a bridging role for F-EP segments on both sides of the filler, which will greatly reduce the interfacial thermal resistance during heat conduction, thus forming an effective heat conduction network and improving the overall thermal conductivity of the composites. At the same time, the interconnected network structure formed between FGO and matrix relies on strong interaction force, and evenly distributed adsorption layers are produced. When the temperature rises, the heat flow can spread more rapidly along the uniform adsorption network structure, thus significantly improving the TC of epoxy composite materials. Based on this, it can be predicted that the thermal conductivity of composite materials can be significantly improved by designing a reasonable spatial structure of the filler network and increasing filler concentration.

3.3. Microscopic Parameter Calculation and Principle Discussion

In order to analyze the influence mechanism of the above performance changes, the free volume, mean square displacement (MSD), and axial density distribution were further calculated.

3.3.1. Free Volume

According to the free volume theory, the total volume (V_t) of solid or liquid substances can be divided into occupied volume (V_0) and free volume (V_f) [48]. The size of the free volume in polymer composites has a significant impact on the thermal and mechanical properties of the material. By analyzing the distribution of free volume inside the material, the mechanism of the change in the macroscopic properties of the material can be explored. In this study, due to the MD treatment during model construction, the volume of different epoxy systems is not exactly the same. Therefore, the V_f of each epoxy system cannot be directly compared. Consequently, the percentage is used to calculate the fractional free volume (FFV) to characterize; its expression is [49]:

$$FFV = \frac{V_f}{V_0 + V_f} \times 100\% \tag{8}$$

This paper calculated the FFV of FGO/F-EP system with different interlayer spacing at 300 K, and the results are shown in Figure 8. With the continuous increase of the distance between the FGO filler layers, the FFV shows a trend of first decreasing and then increasing. Compared with the random system, when the interlayer spacing is 3 Å to 9 Å, the FFV of the composite material is relatively reduced, and when the interlayer spacing is further increased, the FFV of the system increases significantly. We think that FGO with random distribution is easy to agglomerate in the system, which makes the free volume of the system relatively large, while the FGO with reasonable interlayer spacing will be evenly filled in the epoxy system, and the binding effect of FGO on both sides of the matrix material will lead to the orderly arrangement of epoxy molecules, which will reduce the FFV of the system. When the layer spacing is further increased, FGO is distributed too loosely in the composite material, and the binding effect on the epoxy resin molecules is relatively weakened. At the same time, more holes are generated in the interface area, which makes the FFV significantly increased.

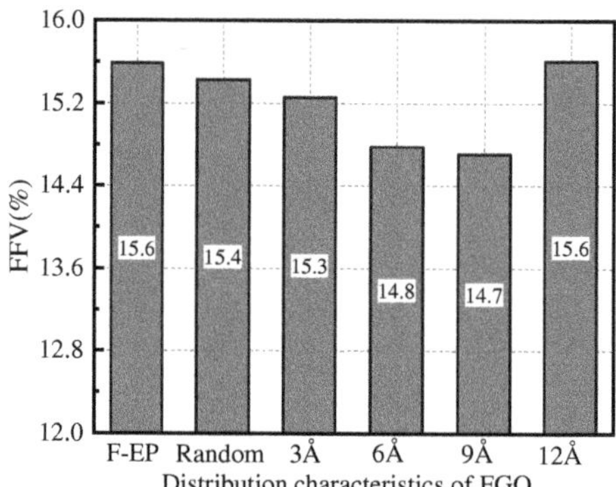

Figure 8. FFV of FGO/F-EP composites with different distribution characteristics.

3.3.2. Mean Square Displacement

A large number of studies have shown that the movement of molecular segments in polymer composites is one of the important factors affecting material properties. The strength of segment movement is directly related to the degree of looseness of the composite network structure. In many application scenarios, reducing the movement capacity of the chain segment is a key method to ensure that the material has high mechanical properties. The strength of the internal chain segment movement of the material can be characterized by the MSD parameter [50], and the MSD is defined as:

$$MSD = \frac{1}{3N} \sum_{i=0}^{N-1} \left(|R_i(t) - R_i(0)|^2 \right) \tag{9}$$

where N is the total number of atoms in the system, $R_i(t)$ and $R_i(0)$ respectively represent the displacement vector of the atom i in the epoxy system at time t and the initial time. This paper calculated the MSD parameters of the FGO/F-EP system with different interlayer spacing at 300 K, and the results are shown in Figure 9. It can be seen from the figure that the order of MSD of F-EP with different FGO distribution is 9 Å < 12 Å < Random < 6 Å < 3 Å, and this result is generally consistent with the trend of the FFV. The analysis suggests that the filler itself has an inhibitory effect on the movement of the epoxy resin molecular chain, but the excessively agglomerated filler has little effect on the matrix, which causes its effect on the epoxy molecular chain to be limited. After FGO is filled in order, a binding layer can be generated in the system through interaction force. At the same time, the fluorine-containing group can also bond with the epoxy matrix. The strong polar C-F bond also forms a certain barrier effect, and further guides the regular distribution of the resin matrix chain segments, thereby effectively reducing the MSD of the system.

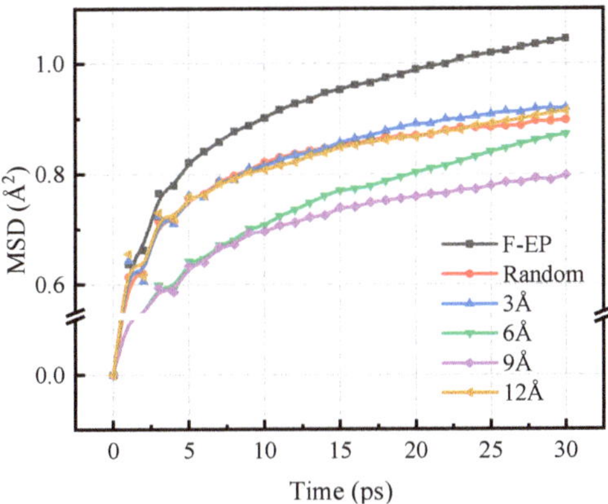

Figure 9. MSD of F-EP composite systems with different FGO layer spacing.

3.3.3. Axial Density Distribution

In order to study the effect of graphene addition on the distribution of resin matrix molecules in the composite material, we calculated the relative density distribution curve of the FGO/F-EP composite material model in the axial direction (perpendicular to the FGO surface). The long distance of epoxy resin along the z-axis is divided into several small areas. The results of axial density distribution are shown in Figure 10.

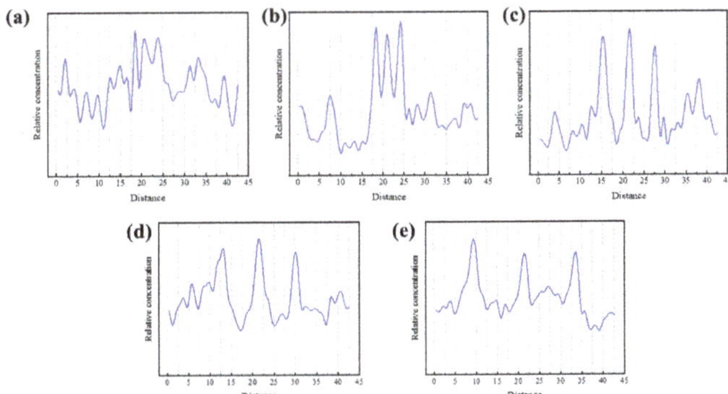

Figure 10. Axial density distribution of FGO/F-EP composites with different distribution characteristics. (**a**) Random; (**b**) 3-FGO/F-EP; (**c**) 6-FGO/F-EP; (**d**) 9-FGO/F-EP; (**e**) 12-FGO/F-EP.

When FGO is disordered, the axial density of the system shows strong fluctuation, and there is no obvious density concentration area. When FGO is filled orderly, the axial density distribution curve of the epoxy composite material model shows three sharp peaks, and the peak position is the z-axis coordinate corresponding to the three intercalated FGO in the system. It can be found that the density of the system near the FGO nanosheets is relatively high, and with the increase of the interlayer spacing, the peak width gradually increases, which indicates that the epoxy resin molecules during the cross-linking reaction will be bound on both sides of the FGO. This is also in line with the previous forecast. The 3 Å model can clearly see the overlap of the peaks, which proves the effect of filler agglomeration. The overlap effect of the 6 Å model is weakened, but the peak width does not change significantly. The peak width of the 9 Å model is significantly increased, and the adsorption effect of the FGO nanosheets on the epoxy molecules can be clearly seen. When the interlayer spacing is further increased, the peak amplitude and peak width in the axial density curve decrease, which is believed to be the reason that the excessive distance between the FGO layers leads to the weakening of the binding effect on the epoxy resin molecules.

Based on the research results, we believe that the pre-fixed two-dimensional filler network can be used to control the cross-linking network of epoxy resin materials. After functionalizing graphene by fluorination, the epoxy resin can be induced to cross-link into bonds. The network structure of epoxy resin can be controlled by the binding effect of the filler network, so that the regular cross-linking region can grow orderly along the spatial structure of filler network. Through this idea, the number of free segments in polymer materials can be effectively reduced, and the movement ability of the segments can be reduced, so as to improve the mechanical strength of the material and increase the glass transition temperature. This will greatly expand the application of polymer composites and provide theoretical guidance for the design of high-performance polymer nanocomposites.

4. Conclusions

In this paper, the model of ordered filled FGO nanosheets was constructed, and the effect of FGO nanosheets spacing on the thermal and mechanical properties of fluorinated epoxy resin composites was studied. It shows that the orderly filling of FGO can significantly improve the static elastic modulus, T_g, CTE, and TC of the composite materials. The interlayer spacing distribution of FGO nanoflakes in the matrix also has an obvious effect on the thermodynamic properties of the composites. The results show that the composite has the best comprehensive properties when the interlayer spacing of FGO is about 9 Å. Among them, Young's modulus, bulk modulus, and shear modulus are increased by 9.2%, 6.36%, and 0.57%, respectively. The glass transition temperature increases by 21.79 K, the thermal conductivity increases by 3.47%, and the thermal expansion coefficient of glass state decreases by 20%.

Furthermore, the influence mechanism of FGO interlayer distance on the properties of the composites was analyzed by calculating the microscopic parameters of the systems. After ordered filling, the aggregation characteristics of FGO are weakened, and the fillers can be evenly distributed in the composites. There is a strong interaction between the fluorinated FGO and F-EP, which can induce the ordered distribution of epoxy resin molecules on both sides of the two-dimensional filler and cross-linking. This makes the FFV of the materials significantly reduced, and the chains segment motion capacity is also limited by the filler network. When the FGO layer spacing is too large, the binding effect is weakened, which leads to the disorder of the microstructure of the composites. By analyzing the axial density distribution of the composite material, it is found that the fixed-space functionalized filler network structure does have an adsorption effect on epoxy resin molecules, and can bind them on both sides of the two-dimensional filler to guide its orderly cross-linking. We predict that there is a key connection between the crystallization behavior and glass transition behavior of epoxy resin materials. In the future, we can control the cross-link behavior of polymer materials by designing functionalized two-dimensional filler networks and develop high-performance polymers composite materials.

Author Contributions: Conceptualization, J.X., Q.D., and Q.X.; methodology, J.X. and Q.D.; software, X.Y. and C.X.; validation, J.X., G.X., and Q.D.; formal analysis, X.Y.; investigation, Q.D. and X.Y.; resources, Q.X. and Z.H.; data curation, X.Y.; writing—original draft preparation, X.Y. and Q.D.; writing—review and editing, Q.D.; visualization, X.Y.; supervision, J.X.; project administration, Q.X.; funding acquisition, J.X. All authors have read and agreed to the published version of the manuscript.

Funding: This research was funded by the National Natural Science Foundation of China, grant number 52007065, and the Self-Topic Fund of State Key Laboratory of Alternate Electrical Power System with Renewable Energy Sources, grant number LAPS202116.

Data Availability Statement: Data is contained within the article.

Conflicts of Interest: The authors declare no conflict of interest.

References

1. Thomas, R.; Ding, Y.M.; He, Y.L.; Yang, L.; Moldenaers, P.; Yang, W.M.; Czigany, T.; Thomas, S. Miscibility, morphology, thermal, and mechanical properties of a DGEBA based epoxy resin toughened with a liquid rubber. *Polymer* **2008**, *49*, 278–294. [CrossRef]
2. Gu, J.W.; Liang, C.B.; Zhao, X.M.; Gan, B.; Qiu, H.; Guo, Y.Q.; Yang, X.T.; Zhang, Q.Y.; Wang, D.Y. Highly thermally conductive flame-retardant epoxy nanocomposites with reduced ignitability and excellent electrical conductivities. *Compos. Sci. Technol.* **2017**, *139*, 83–89. [CrossRef]
3. Zhao, Y.S.; He, Y.H.; Yang, K.R.; Wang, X.P.; Bai, J.H.; Du, B. Improving the Surface Insulating Performance of Epoxy Resin/Al$_2$O$_3$ Composite Materials by Extending Chain of Liquid Epoxy Resin with Me-THPA. *High Volt.* **2019**, *5*, 472–481. [CrossRef]
4. Xie, Q.; Fu, K.X.; Liang, S.D.; Liu, B.W.; Lu, L.; Yang, X.M.; Huang, Z.Y.; Lu, F.C. Micro-Structure and Thermomechanical Properties of Crosslinked Epoxy Composite Modified by Nano-SiO$_2$: A Molecular Dynamics Simulation. *Polymer* **2018**, *10*, 801. [CrossRef]
5. Barbosa, A.P.C.; Fulco, A.P.P.; Guerra, E.S.S.; Arakaki, F.K.; Tosatto, M.; Costa, M.C.B.; Melo, J.D.D. Accelerated aging effects on carbon fiber/epoxy composites. *Compos. Part B Eng.* **2016**, *110*, 298–306. [CrossRef]
6. Mondal, N.; Haque, N.; Dalai, S.; Chakravorti, S.; Chatterjee, B. A Method for Identifying Ageing in Epoxy-mica Composite Insulation used in Rotational machines through Modelling of Dielectric Relaxation. *High Volt.* **2020**, *5*, 184–190. [CrossRef]
7. Brostow, W.; Chonkaew, W.; Menard, K.P.; Scharf, T.W. Modification of an epoxy resin with a fluoroepoxy oligomer for improved mechanical and tribological properties. *Mat. Sci. Eng.* **2009**, *507*, 241–251. [CrossRef]
8. Hou, D.F.; Ma, H.B.; Li, X.Y.; He, J.; Liao, H.W. Preparation and properties of 4,4'-(hexafluoroisopropylidene) diphenol cured 2,2-bis(4-cyanatophenyl) propane. *J. Appl. Polym. Sci.* **2017**, *134*, 44518. [CrossRef]
9. Taa, Z.Q.; Yang, S.Y.; Ge, Z.Y.; Chen, J.S.; Fan, L. Synthesis and properties of novel fluorinated epoxy resins based on 1,1-bis(4-glycidylesterphenyl)-1-(3'-trifluoromethylphenyl)-2,2,2-trifluoroethane. *Eur. Polym. J.* **2007**, *43*, 550–560. [CrossRef]
10. Wan, Y.J.; Tang, L.C.; Gong, L.X.; Yan, D.; Li, Y.B.; Wu, L.B.; Jiang, J.X.; Lai, G.Q. Grafting of epoxy chains onto graphene oxide for epoxy composites with improved mechanical and thermal properties. *Carbon* **2014**, *69*, 467–480. [CrossRef]
11. Shirasu, K.; Kitayama, S.; Liu, F.; Yamamoto, G.; Hashida, T. Molecular Dynamics Simulations and Theoretical Model for Engineering Tensile Properties of Single-and Multi-Walled Carbon Nanotubes. *Nanomaterials* **2021**, *11*, 795. [CrossRef] [PubMed]
12. Huang, T.; Lu, R.G.; Su, C.; Wang, H.N.; Guo, Z.; Liu, P.; Huang, Z.Y.; Chen, H.M.; Li, T.S. Chemically modified graphene/polyimide composite films based on utilization of covalent bonding and oriented distribution. *ACS Appl. Mater. Interfaces* **2012**, *4*, 2699–2708. [CrossRef] [PubMed]
13. Fang, F.; Ran, S.Y.; Fang, Z.P.; Song, P.G.; Wang, H. Improved flame resistance and thermo-mechanical properties of epoxy resin nanocomposites from functionalized graphene oxide via self-assembly in water. *Composites* **2019**, *165*, 406–416. [CrossRef]
14. McAllister, M.J.; Li, J.L.; Adamson, D.H.; Schniepp, H.C.; Abdala, A.A.; Liu, J.; Herrera-Alonso, M.; Milius, D.L.; Car, R.; Prud'homme, R.K.; et al. Single sheet functionalized graphene by oxidation and thermal expansion of graphite. *Chem. Mater.* **2007**, *19*, 4396–4404. [CrossRef]
15. Kim, H.; Miura, Y.; Macosko, C.W. Graphene/Polyurethane Nanocomposites for Improved Gas Barrier and Electrical Conductivity. *Chem. Mater.* **2010**, *22*, 3441–3450. [CrossRef]
16. Zhao, Y.; Hu, Z. Graphene in Ionic Liquids: Collective van der Waals Interaction and Hindrance of Self-Assembly Pathway. *J. Phys. Chem. B* **2013**, *117*, 10540–10547. [CrossRef]
17. Lin, S.; Shih, C.J.; Strano, M.S.; Blankschtein, D. Molecular Insights into the Surface Morphology, Layering Structure, and Aggregation Kinetics of Surfactant-Stabilized Graphene Dispersions. *J. Am. Chem. Soc.* **2011**, *133*, 12810–12823. [CrossRef] [PubMed]
18. Lü, F.; Zhan, Z.; Zhang, G.; Jiao, Y.; Xie, Q. Effect of Plasma Fluorinated Modified Micron AlN Filling on Insulation Properties of Epoxy Resin. *Trans. China Electrotech. Soc.* **2019**, *34*, 3522–3531.
19. Wu, H.; Xie, Q.; Duan, Q.; Yan, J.; Yin, K. Plasma fluorination of BaTiO$_3$ for enhancement ofinterfacial adhesion and surface insulation of epoxy resin. *J. Mater. Sci.* **2020**, *4*, 1499–1510. [CrossRef]
20. Lu, F.; Ruan, H.; Song, J.; Yin, K.; Zhan, Z.; Jiao, Y.; Xie, Q. Enhanced surface insulation and depressed dielectricconstant for Al$_2$O$_3$/epoxy composites through plasma fluorination of filler. *J. Phys. D Appl. Phys.* **2019**, *52*, 155201. [CrossRef]
21. Yang, X.; Yu, D.; Cao, B.; To, A. Ultrahigh Thermal Rectification in Pillared Graphene Structure with Carbon Nanotube–Graphene Intramolecular Junctions. *ACS Appl. Mater. Interfaces* **2017**, *9*, 29–35. [CrossRef]
22. Wang, Y.; Yang, C.; Pei, Q.; Zhang, Y. Some Aspects of Thermal Transport across the Interface between Graphene and Epoxy in Nanocomposites. *ACS Appl. Mater. Interfaces* **2016**, *8*, 8272–8279. [CrossRef] [PubMed]
23. Chen, H.; Ginzburg, V.V.; Yang, J.; Yang, Y.; Liu, W.; Huang, Y.; Du, L.; Chen, B. Thermal Conductivity of Polymer-Based Composites: Fundamentals and Applications. *Prog. Polym. Sci.* **2016**, *59*, 41–85. [CrossRef]

24. Yousefi, N.; Sun, X.; Lin, X.; Shen, X.; Jia, J.; Zhang, B.; Tang, B.; Chan, M.; Kim, J.K. Highly Aligned Graphene/Polymer Nanocomposites with Excellent Dielectric Properties for High Performance Electromagnetic Interference Shielding. *Adv. Mater.* **2015**, *26*, 5480–5487. [CrossRef] [PubMed]

25. Fang, M.; Hao, Y.; Ying, Z.; Wang, H.; Cheng, H.; Zeng, Y. Controllable edge modification of multi-layer graphene for improved dispersion stability and high electrical conductivity. *Appl. Nanosci.* **2019**, *9*, 469–477. [CrossRef]

26. Xiao, C.; Guo, Y.; Tang, Y.; Ding, J.; Zhang, X.; Zheng, K.; Tian, X. Epoxy composite with significantly improved thermal conductivity by constructing a vertically aligned three-dimensional network of silicon carbide nanowires/boron nitride nanosheets. *Compos. Part B Eng.* **2020**, *187*, 107855. [CrossRef]

27. Kumar, S.K.; Ganesan, V.; Riggleman, R.A. Perspective: Outstanding theoretical questions in polymer-nanoparticle hybrids. *J. Chem. Phys.* **2017**, *147*, 020901. [CrossRef]

28. Jabbarzadeh, A.; Halfina, B. Unravelling the effects of size, volume fraction and shape of nanoparticle additives on crystallization of nanocomposite polymers. *Nanoscale Adv.* **2019**, *1*, 4704. [CrossRef]

29. Jabbarzadeh, A. The Origins of Enhanced and Retarded Crystallization in Nanocomposite Polymers. *Nanomaterials* **2019**, *9*, 1472. [CrossRef]

30. Yousefi, N.; Gudarzi, M.M.; Zheng, Q.; Aboutalebi, S.H.; Sharif, F.; Kim, J.K. Self-alignment and high electrical conductivity of ultralarge graphene oxide–polyurethane nanocomposites. *J. Mater. Chem. A* **2012**, *22*, 12709–12717. [CrossRef]

31. Yousefi, N.; Gudarzi, M.M.; Zheng, Q.; Lin, X.; Shen, X.; Jia, J.; Sharif, F.; Kim, J.K. Highly aligned, ultralarge-size reduced graphene oxide/polyurethane nanocomposites: Mechanical properties and moisture permeability. *Compos. Part A Appl. Sic. Manuf.* **2013**, *49*, 42–50. [CrossRef]

32. Li, Q.; Huang, X.; Liu, T.; Yan, J.; Wang, Z.; Zhang, Y.; Lu, X. Advances in application of molecular simulation technology in high voltage insulation. *Trans. China Electrotech. Soc.* **2016**, *31*, 1–13.

33. Yang, Q.; Li, X.; Shi, L.; Yang, X.; Sui, G. The thermal characteristics of epoxy resin: Design and predict by using molecular simulation method. *Polymer* **2013**, *54*, 6447–6454. [CrossRef]

34. Zhang, X.; Chen, X.; Xiao, S.; Wen, H.; Wu, Y. Molecular dynamics simulation of thermodynamic properties of modified SiO_2 epoxy resin. *High Volt. Technol.* **2018**, *44*, 740–749.

35. Damasceno, D.A.; Rajapakse, R.; Mesquita, E. Atomistic Modelling of Size-Dependent Mechanical Properties and Fracture of Pristine and Defective Cove-Edged Graphene Nanoribbons. *Nanomaterials* **2020**, *10*, 1422. [CrossRef]

36. Chowdhury, S.C.; Prosser, R.; Sirk, T.W.; Elder, R.M.; Gillespie, J.W. Glass fiber-epoxy interactions in the presence of silane: A molecular dynamics study. *Appl. Surf. Sci.* **2020**, *542*, 148738. [CrossRef]

37. Zhang, W.; Qing, Y.; Zhong, W.; Sui, G.; Yang, X.P. Mechanism of modulus improvement for epoxy resin matrices: A molecular dynamics simulation. *React. Funct. Polym.* **2017**, *111*, 60–67. [CrossRef]

38. Hadden, C.M.; Klimek-Mcdonald, D.R.; Pineda, E.J.; King, J.A.; Reichanadter, A.M.; Miskioglu, I.; Gowtham, S.; Odegard, G.M. Mechanical properties of graphene nanoplatelet/carbon fiber/epoxy hybrid composites: Multiscale modeling and experiments. *Carbon* **2015**, *95*, 100–112. [CrossRef]

39. Shenogina, N.B.; Tsige, M.; Patnaik, S.S.; Mukhopadhyay, S.M. Molecular Modeling Approach to Prediction of Thermo-Mechanical Behavior of Thermoset Polymer Networks. *Macromolecules* **2012**, *45*, 5307–5315. [CrossRef]

40. Wang, T.Y.; Tseng, P.Y.; Tasi, J.L. Characterization of Young's modulus and thermal conductivity of graphene/epoxy nanocomposites. *J. Compos. Mater.* **2019**, *53*, 835–847. [CrossRef]

41. Yang, K.R.; Zhao, Y.S.; Zhang, S.; He, Y.H.; Wang, X.P. Study on Thermal and Electrical Properties of BPAF Modified Epoxy/Al_2O_3 Composites. In Proceedings of the 2019 2nd International Conference on Electrical Materials and Power Equipment (ICEMPE), Guangzhou, China, 7–10 April 2019; pp. 277–280.

42. Tersoff, J. Modeling solid-state chemistry: Interatomic potentials for multicomponent systems. *Phys. Rev. B Condens. Matter.* **1989**, *39*, 5566–5568. [CrossRef] [PubMed]

43. Shokuhfar, A.; Arab, B. The effect of cross linking density on the mechanical properties and structure of the epoxy polymers: Molecular dynamics simulation. *J. Mol. Modeling* **2013**, *19*, 3719–3731. [CrossRef]

44. Jeyranpour, F.; Alahyarizadeh, G.; Minuchehr, A. The thermo-mechanical properties estimation of fullerene-reinforced resin epoxy composites by molecular dynamics simulation–A comparative study. *Polymer* **2016**, *88*, 9–18. [CrossRef]

45. Choi, J.; Yu, S.; Yang, S.; Cho, M. The glass transition and thermoelastic behavior of epoxy-based nanocomposites: A molecular dynamics study. *Polymer* **2011**, *52*, 5197–5203. [CrossRef]

46. Müller-Plathe, F. A simple nonequilibrium molecular dynamics method for calculating the thermal conductivity. *J. Chem. Phys.* **1997**, *106*, 6082–6085. [CrossRef]

47. Li, S.H.; Yu, X.X.; Bao, H.; Yang, N. High thermal conductivity of bulk epoxy resin by bottom-up parallel-linking and strain: A molecular dynamics study. *Indian J. Chem. A* **2018**, *122*, 13140–13147. [CrossRef]

48. Fox, T.G. Second-order transition temperatures and related properties of polystyrene. I. influence of molecular weight. *J. Appl. Phys.* **1950**, *6*, 581–591. [CrossRef]

49. Wei, Q.H.; Zhang, Y.F.; Wang, Y.E.; Yang, M.M. A molecular dynamic simulation method to elucidate the interaction mechanism of nano-SiO_2 in polymer blends. *J. Mater. Sci.* **2017**, *52*, 12889–12901. [CrossRef]

50. Li, K.; Li, Y.; Lian, Q.S.; Cheng, J.; Zhang, J.Y. Influence of cross-linking density on the structure and properties of the interphase within supported ultrathin epoxy films. *J. Mater. Sci.* **2016**, *51*, 9019–9030. [CrossRef]

Article

Fabrication of Pressure Sensor Using Electrospinning Method for Robotic Tactile Sensing Application

Tamil Selvan Ramadoss [1,*], Yuya Ishii [2], Amutha Chinnappan [1], Marcelo H. Ang [1] and Seeram Ramakrishna [1]

[1] Department of Mechanical Engineering, Faculty of Engineering, National University of Singapore, 9 Engineering Drive 1, Singapore 117575, Singapore; mpecam@nus.edu.sg (A.C.); mpeangh@nus.edu.sg (M.H.A.); seeram@nus.edu.sg (S.R.)

[2] Faculty of Fiber Science and Engineering, Kyoto Institute of Technology, Kyoto 606-8585, Japan; yishii@kit.ac.jp

* Correspondence: mperts@nus.edu.sg

Abstract: Tactile sensors are widely used by the robotics industries over decades to measure force or pressure produced by external stimuli. Piezoelectric-based pressure sensors have intensively been investigated as promising candidates for tactile sensing applications. In contrast, piezoelectric-based pressure sensors are expensive due to their high cost of manufacturing and expensive base materials. Recently, an effect similar to the piezoelectric effect has been identified in non-piezoelectric polymers such as poly(d,l-lactic acid (PDLLA), poly(methyl methacrylate) (PMMA) and polystyrene. Hence investigations were conducted on alternative materials to find their suitability. In this article, we used inexpensive atactic polystyrene (aPS) as the base polymer and fabricated functional fibers using an electrospinning method. Fiber morphologies were studied using a field-emission scanning electron microscope and proposed a unique pressure sensor fabrication method. A fabricated pressure sensor was subjected to different pressures and corresponding electrical and mechanical characteristics were analyzed. An open circuit voltage of 3.1 V was generated at 19.9 kPa applied pressure, followed by an integral output charge (ΔQ), which was measured to calculate the average apparent piezoelectric constant d_{app} and was found to be 12.9 ± 1.8 pC N^{-1}. A fabricated pressure sensor was attached to a commercially available robotic arm to mimic the tactile sensing.

Keywords: electrospinning; non-piezoelectric polymers; tactile sensors; robotic gripper

Citation: Ramadoss, T.S.; Ishii, Y.; Chinnappan, A.; Ang, M.H.; Ramakrishna, S. Fabrication of Pressure Sensor Using Electrospinning Method for Robotic Tactile Sensing Application. *Nanomaterials* **2021**, *11*, 1320. https://doi.org/10.3390/nano11051320

Academic Editors: Rossana Dimitri and Francesco Tornabene

Received: 16 April 2021
Accepted: 13 May 2021
Published: 17 May 2021

Publisher's Note: MDPI stays neutral with regard to jurisdictional claims in published maps and institutional affiliations.

1. Introduction

Tactile sensing technology has undoubtedly attracted more attention in the field of robotics. These sensors have been looked upon as the best choice for improving the precision and contact with a robot's environment [1–3]. Tactile sensing devices, as shown in big 1a, which emulate the properties of human skin, in other words electronic skin structures, have been widely investigated for use in various practical applications such as smart prosthetics, [4] wearable devices, [5] and artificial robot arms [6,7]. For robotic applications, the two fundamental research areas on tactile sensor development connected to object-controlled lifting and grasp functioning with the ability to describe diverse surface textures have developed continuously [8].

In the last decade, various types of pressure sensors, including resistive, [9] capacitive, [10] triboelectric [11] and piezoelectric [12] sensors, have been intensively examined as promising candidates for tactile sensing applications. Among all sensor types, piezoelectric-based pressure sensors are autonomously powered, which generate a consistent electrical signal in response to applied force/pressure. Hence the piezoelectric-type sensor represents a prominent, stand-alone and self-reliant sensor, which makes it suitable for robotic tactile sensing to provide reliable electrical inputs for robotic manipulators [13].

With the development of materials science and versatile fabrication techniques, two-dimensional (2D) and three-dimensional (3D) resilient nanostructures have been synthesized sophisticatedly and are receiving greater attention among researchers [14]. In this direction, electrospinning fabrication methods as shown in Figure 1c, have received significant consideration to produce polymer-based one-dimensional functional nanofibers, as shown in Figure 1d [15,16].

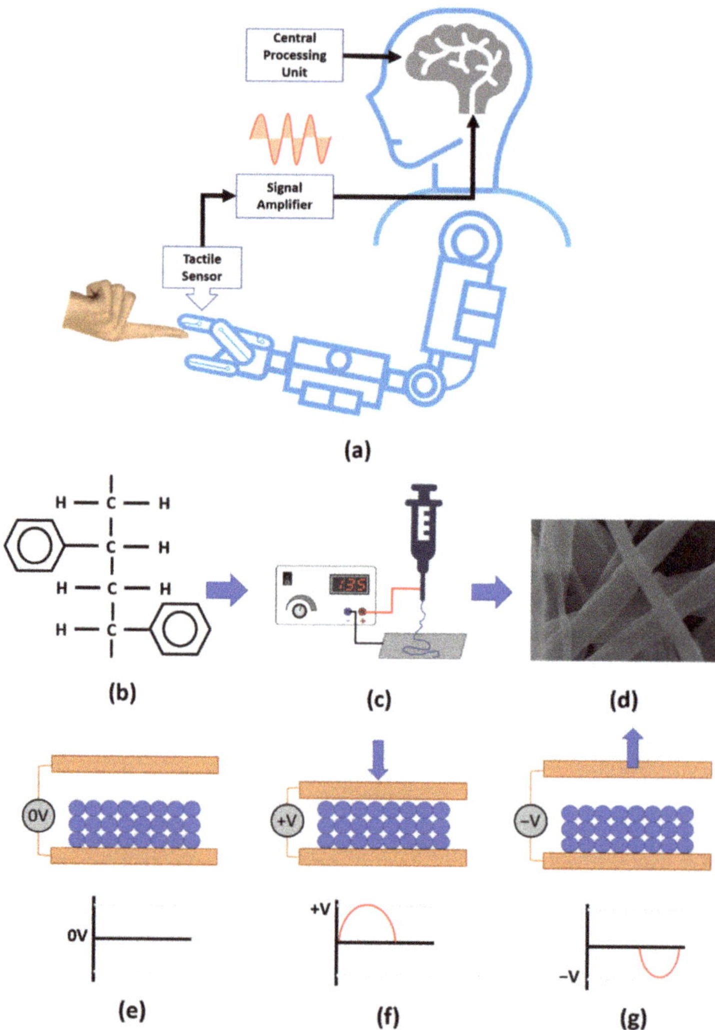

Figure 1. (**a**) Schematic illustration of a robot touch sensation. (**b**) Atactic polystyrene chain. (**c**) Electrospinning set-up. (**d**) SEM image of nanofibers. Schematics of tactile sensor (**e**) neutral, (**f**) compressed and (**g**) released.

Micrometer/sub micrometer electromechanical polymer fibers are promising components for tactile sensing application due their high mechanical flexibility, low weight, and excellent breathability [17]. These polymer-based sensors primarily comprise piezoelectric polymers, including poly(vinylidene fluoride) (PVDF) [18], poly(vinylidenefluoride-co-trifluoroethylene) (PVDF-TrFE) [19] and Barium titanite [20]. Although piezoelectric polymer fibers have already been demonstrated as pressure sensors by various research groups, piezoelectric polymers are expensive. Hence, there are challenges for low-cost manufacturing and large-area fabrication which limits the practical application. On the other hand, converse electromechanical responses were observed inadvertently with electrospun submicron/micron fiber mats composed of nonpiezoelectric polymers, such as poly(D,L-lactic acid (PDLLA), [21] poly(methyl methacrylate) (PMMA) [21], and their composites [22]. These non-piezoelectric materials exhibit piezoelectric properties like conventional piezoelectric materials, indicating a significant high apparent piezoelectric d constant $\left(d_{app}\right)$, as high as 8500 pm V^{-1} for an individual material [21] and 29,000 pm V^{-1} for a composite material [22]. These papers also reported that these excellent electromechanical properties are partly or mainly attributed to the unique electrically charged and mechanically soft nature of the electrospun fiber mats. Observing piezoelectric properties in non-piezoelectric material is a fairly new concept, for instance, D. Hassan, et al. demonstrated piezoelectricity using polystyrene-copper oxide (PS-CuO) nano composites. The study suggested that the (PS-CuO) nanocomposites are highly sensitive for pressure and the electrical resistance of nanocomposites decreases with the increase in pressure [23]. Similarly, Y. Ishii, et al. demonstrated the piezoelectric properties in polystyrene polymers (Figure 1b) and explored the origin of the piezoelectric response. The apparent piezoelectric constant of the fiber mat was measured as 950–1400 pC/N with an applied load of 0.05–0.28 N using the quasistatic method [24]. Conventionally constructing piezoelectric sensors using polymers/composites or crystals seems to be expensive due to base material and fabrication cost. Meanwhile, polystyrene holds a promising future due to its inexpensive base material, low cost fabrication method and large area fabrication. Figure 1e–g represents the device-operating mechanism; Figure 1e denotes that the sensor has zero output potential at an ideal state. Figure 1f shows that the sensor exhibits a positive potential at the terminals when a force is applied. Similarly, Figure 1g represents a negative potential at the electrode terminal when force is removed from the sensor. In this work, we have fabricated atactic polystyrene (aPS) microfibers using an electrospinning method. An as-spun aPS fiber mat was characterized using a field-emission scanning electron microscope. Subsequently, we constructed a pressure sensor using aPS microfibers; electrical and mechanical characteristics were studied. Finally, the pressure sensor was attached to a commercially available Kuka robotic arm structure for tactile sensing application.

2. Materials and Methods

2.1. Materials and Fabrication of Fibers

Atactic polystyrene (aPS) was chosen as the base material of electrospun fibers because it is a widely used and inexpensive nonpiezoelectric polymer. aPS ($M_w \approx 280,000$, Sigma-Aldrich, Saint Louis, Missouri, USA,) was dissolved in N,N-dimethylformamide (DMF) with a concentration of 30 wt% at room temperature. Next, the solution was electrospun with commercial apparatus (NANON, Mecc, Ogari-shi, Fukuoka, Japan). Details of the electrospinning conditions are as follows: the solution was fed with a constant rate of 5.0 mL h^{-1} and discharged through a single stainless steel needle (inner diameter of 0.59 mm); here, the needle moved reciprocally in a straight line with a width and speed of 60 mm and 10 mm s^{-1}, respectively, to produce the electrospun fibers in a wide area. A metallic drum with a diameter and width of 210 mm and 660 mm, respectively, was placed 17 cm below the tip of the needle to support the following collector sheet on which the electrospun aPS fibers were deposited. The collector sheet was composed of a thin polypropylene (PP) sheet (25 mm $\times$ 310 mm with a thickness of 0.15 mm), on one side of which a conductive carbon tape (20 mm $\times$ 310 mm with a thickness of 0.11 mm) was attached to enhance the adhesion of the electrospun fibers as shown Figure 2a. The collector sheet was attached to the surface of the drum; then, the carbon tape and drum had electrically connected to each other and were grounded; 12.0 kV was applied to the needle and the electrospun aPS fibers were deposited for 20 min. The rotating speed of the drum was 50 rpm.

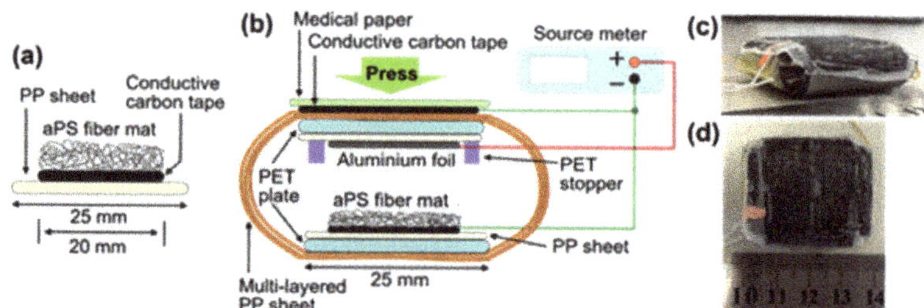

Figure 2. Schematics of (**a**) the collector sheet and (**b**) developed pressure sensor. Photographs of the sensor: (**c**) side-view and (**d**) top view.

2.2. Sensor Structure

The schematic and pictures of the sensor developed in this study is shown in Figure 2b–d. Its operation mechanism as a pressure sensor is basically explained using the electret condenser model reported recently [17]. Briefly, the as-electrospun aPS fiber was deposited on the bottom electrode, which is the conductive carbon tape in the present study, uniquely holds space charges with both positive and negative polarities [24]. In addition, the positive and negative space charges are generally distributed in the upper and lower parts of the fiber mat, respectively. These unique space charges induce compensation charges in the bottom and upper electrodes. When the distance between the bottom and upper electrodes change due to application of an external pressure to the sensor and a resulting deformation of the sensor, the amount of the space charges in a steady state change. This change in the charge amount in the steady state outputs charges so that pressing and releasing the sensor generate charges and output voltage.

The collector sheet, on which the aPS fiber mat was deposited, cut to the size of 25 mm × 30 mm and attached to the PET plate (25 mm × 30 mm with a thickness of 0.7 mm) and used as the bottom part as shown in Figure 2b. An aluminum foil (20 mm × 30 mm with a thickness of 0.1 mm) was used as a upper electrode, which was attached on a PP sheet (25 mm × 30 mm with a thickness of 0.15 mm) and PET plate (25 mm × 30 mm with a thickness of 0.7 mm). The multi-layered PP sheet in Figure 2b was composed of thin PP sheets with the width and thickness of 25 mm and 0.15 mm, respectively. The multi-layered sheet kept the upper and bottom electrodes uncontacted and deformed when the external pressure was applied to the sensor. The PET stopper (20 mm/15 mm × 1 mm with a thickness of 0.7 mm) was attached to the upper plate to prevent the upper electrode from hardly touching and damaging the aPS fiber mat on the bottom electrode. The conducting carbon tape on the multi-layered PP sheet and medical paper shown in Figure 2b were introduced to reduce the generation of contacting charges when the sensor contacted to an object and released it.

2.3. Characterization

The shape of the aPS fiber mat was characterized using a field-emission scanning electron microscope (FESEM; S-4300, Hitachi, Marunouchi, Tokyo, Japan) after a 6-nm-thick gold coating. Figure 3a shows the top view of the FESEM image of the as-electrospun aPS fiber mat. The fibers have smooth morphologies without beaded morphologies and orients to random directions. The average diameter is determined to be 5.89 ± 0.39 µm (all errors in this paper represent the standard deviation) from 100 points of diameters measured from the FESEM images. Figure 3b shows the cross-sectional FESEM image of the fiber mat. The aPS fibers stack with air spaces and form a fiber mat. The thickness of the fiber mat is approximately 210 µm. Quantitative testing of the electrical outputs of the sensors was performed using the handmade setup shown in Figure 4. Here, the sensor was attached to the L-shaped metallic holder and then it was lowered with a constant speed of 4.0 mm s^{-1} through an X-Axis dovetail feed screw (XLSL120-2, Misumi, Bunkyo-ku, Tokyo, Japan) and DC motor (HG37-30-AB-00, Nidec Copal Electronics, Tokyo, Japan). The applied load to the sensor was measured with the gravimeter (ACS5000, Kyoto, Japan);

the load was controlled at the same time as lowering the sensor. The applied load was converted to pressure, dividing the load by the bottom area of the sensor, $6.25 \times 10^{-4} \text{ m}^2$. The initial height of the bottom of the sensor was fixed 5 mm from the surface of the gravimeter as shown in Figure 4a. The time-dependent open circuit voltage, $V_{oc}(t)$, and short circuit current, $I_{sc}(t)$, were measured with a source meter (Model 2450, Keithley, Beaverton, Oregon, USA). Here, t represents time.

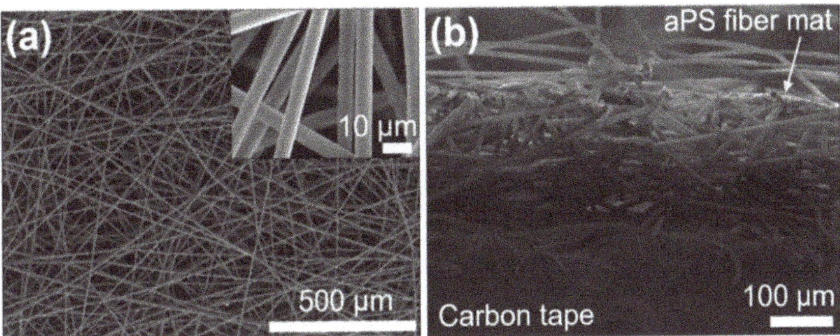

Figure 3. FESEM images of the aPS fiber mat: (**a**) top view and (**b**) cross-section view. Inset in (**a**) shows the enlarged FESEM image of the fiber mat with the top view.

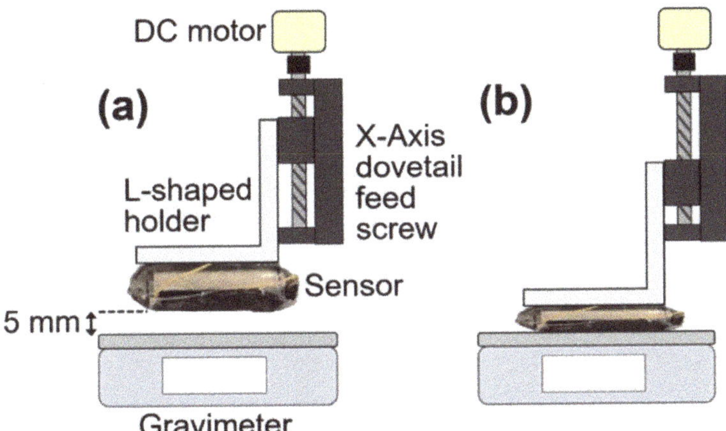

Figure 4. Schematics of the quantitative testing method for the electrical outputs of the sensor: (**a**) the initial state and (**b**) loading state.

3. Results

3.1. Electrical Characteristics

Figure 5a shows $V_{oc}(t)$ from the sensor when the sensor was loaded with different values of pressures. $V_{oc}(t)$ was also measured when the pressure was released, as shown in Figure 5b. The open circuit voltage is generated when the sensor is both loaded and released, which demonstrates that the present sensor acts as a pressure sensor. The absolute values of $V_{oc}(t)$ in each peak increases with increasing the applied pressure, as also plotted in Figure 5c. This result shows that the applied pressure can be identified from the open circuit voltage of the sensor on the condition that the speed for applying pressure is a constant value. The maximum open circuit voltage reaches approximately 3.1 V with the pressure of 19.9 kPa. Although the present study fixed the speed for applying pressure at 4.0 mm s^{-1}, the open circuit voltage can increase with an increase in the speed. The maximum $V_{oc}(t)$ with the applied pressure $\geq$19.9 kPa is saturated because the PET stopper touches the bottom substrate so that the distance between the bottom and the upper electrodes cannot change with the pressure $\geq$19.9 kPa.

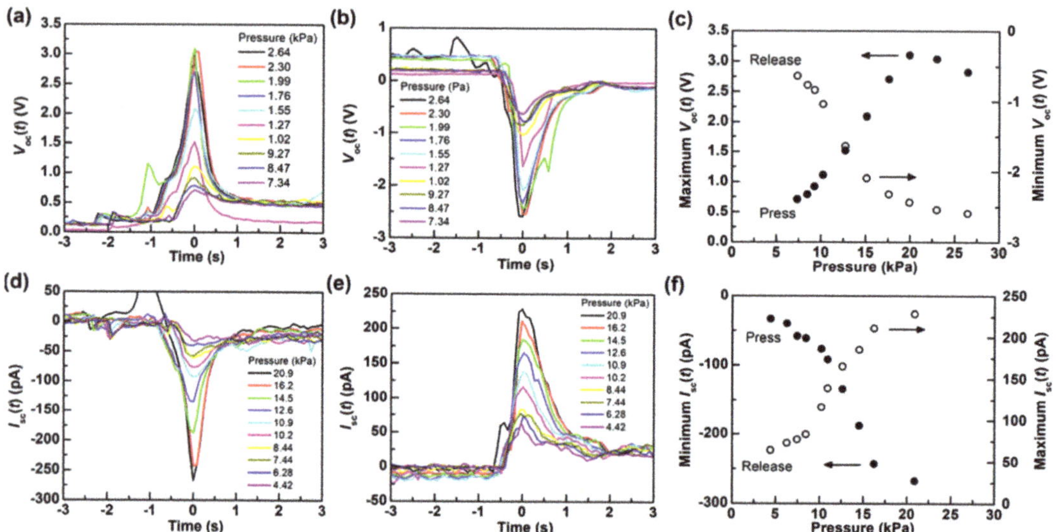

Figure 5. (**a**,**b**) $V_{oc}(t)$ when the sensor was (**a**) loaded and (**b**) released. (**c**) The maximum/minimum $V_{oc}(t)$ with different applied pressures. (**d**,**e**) $I_{sc}(t)$ when the sensor was (**d**) loaded and (**e**) released. (**f**) The maximum/minimum $I_{sc}(t)$ with different applied pressures.

Figure 5d,e show $I_{sc}(t)$ from the sensor when the sensor was loaded and released with/from different values of pressures, respectively. The short circuit current is also generated by loading and releasing the pressure. In addition, the absolute values of $I_{sc}(t)$ in each peak increases with an increase in the applied pressure, as also plotted in Figure 5f, which is similar to the result of the open circuit voltage. The $I_{sc}(t)$ when the sensor was loaded was integrated with the time of loading (ΔT) using the following equation:

$$\int^{\Delta T} I_{sc}(t)\,\mathrm{d}t = \Delta Q \tag{1}$$

Here, ΔQ represents the integrated output charge when the sensor was loaded. ΔQ changes depending on the distance between the bottom and upper electrodes as reported recently [17], such that ΔQ changes depending on the value of the applied pressure in the present sensor. Here, the $V_{oc}(t)$ and $I_{sc}(t)$ from the present sensor changes depending on the speed for applying pressure; however, ΔQ can be constant even when the speed changes. Figure 6a shows ΔQ with different applied pressures. ΔQ increased with the increasing applied pressure. Hence, the applied pressure can be identified from the ΔQ even when the speed for applying the pressure changes.

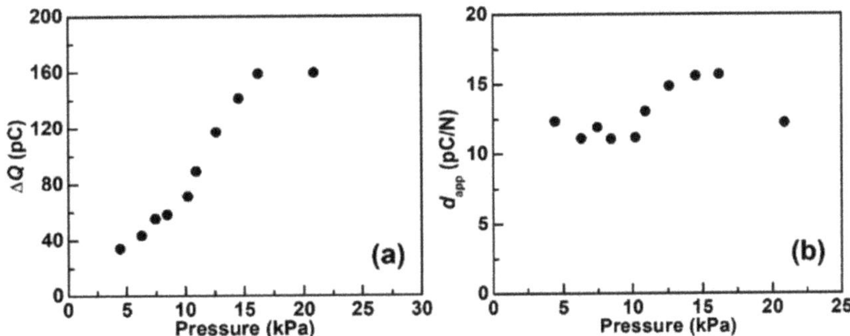

Figure 6. (a) ΔQ with different applied pressures. (b) d_{app} with different applied pressures.

The apparent piezoelectric d constant (d_{app}) was calculated from the following equation:

$$d_{app} = \Delta Q/F \tag{2}$$

where, F is the load applied to the sensor. Figure 6b shows d_{app} with different applied pressures. The d_{app} is roughly constant for the applied pressure with the range from 11.1 pC N^{-1} to 15.7 pC N^{-1}; the average d_{app} is 12.9 ± 1.8 pC N^{-1}. A detailed comparison is presented in the Table 1, representing electromechanical characteristics and fabrication parameters for non-piezoelectric polymers.

Table 1. Comparison of different non piezoelectric materials fabricated via electrospinning and the electromechanical characteristics.

Materials	Average Fiber Size	Electrospinning Parameters		Principle	Electromechanical Properties	Reference
poly(methyl methacrylate) (PMMA)	1 μm	Rate Syringe size Applied Voltage Distance	0.1 mL/h 0.18 mm 8 kV 10 cm	Actuation	Piezoelectric constant (d_T) = 8.5 nm/V	[21]
polymer poly(DL-lactic acid) (PDLLA)	0.4 μm	Rate Syringe size Applied Voltage Distance	0.04 mL/h 0.18 mm 4 kV 10 cm	Actuation	Young modulus = 1.5 kPa Piezoelectric constant = $29,000 \times 10^{-12}$ m V^{-1}	[22]
atactic polystyrene (aPS)	5.8 μm	Rate Syringe size Applied Voltage Distance	5.0 mL/h 0.59 mm 12 kV 17 cm	Sensing	Apparent piezoelectric d constant (d_{app}) = 12.9 ± 1.8 pC N^{-1} Young modulus = 47.7 kPa	This study

3.2. Mechanical Characteristics

The height of the sensor was measured when different pressures were applied to the sensor (Figure 7a). The initial height of the sensor is approximately 10.5 mm; the height is almost maintained up to the applied pressure of 3.4 kPa, which should be considered as the threshold pressure until the sensor starts to deform.

With the pressure range from 4.0 kPa to 17.9 kPa, the height decreases with the increasing applied pressure, which represents the sensor deforms increasing with the pressure. With a pressure of $\geq$19.9 kPa, the height shows an almost constant value, which is because the PET stopper touches the bottom substrate, so that the distance between the bottom and upper electrodes cannot change.

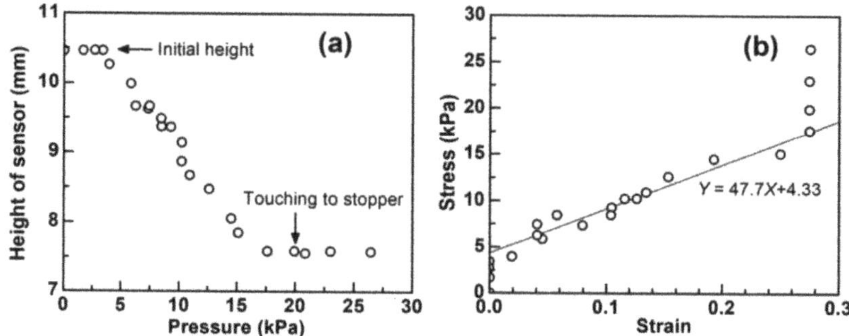

Figure 7. (a) Height of the sensor when a different amount of pressure was applied to the sensor. (b) Stress–strain plots for the sensor.

Figure 7b shows the stress–strain plots for the present sensor. Here, the strain was calculated by dividing the decrement in the height of the sensor by the initial height of the sensor. As observed in Figure 7a, the threshold stress with the stress of ≤3.4 kPa and the upper limit of the strain with the strain of around 0.27 was observed. On the other hand, at the strain range from 0.02 to 0.25, plots are modestly fitted with a linear function. From the tilting of the fitted linear function, the elastic modulus of the sensor was evaluated to be approximately 47.7 kPa. In this study, the multi-layered PP sheet was used to keep the upper and bottom electrodes uncontacted; however, the sheet can be replaced by other materials and structures so that the elastic modulus of the sensor can be simply controlled. The controllable elastic modulus will expand the application fields. In addition, a lower elastic modulus of the sensor with softer structures provides a higher d_{app}.

4. Application

Commercially available tactile sensors such as thin film, capacitive, optical sensors, and magnetic sensors are expensive due to complex fabrication technologies. Moreover, the practical application is inhibited due to the demanding computational power to perform the manipulation process. Meanwhile, the electrospinning fabrication method represents a huge advantage when compared to other fabrication methodologies due to the ease of fabrication and mass production at room temperature. Furthermore, the discovery of piezo-electricity in polystyrene could pave new paths to new research directions and commercial usages. The fabricated pressure sensor (25 × 25 × 7 mm) as shown in Figure 2c,d was connected to a commercially available Kuka LBR IIWA14 R820 model to mimic the tactile sensing application [25]. The Kuka robotic arm was programmed to grab and release an aluminum bar weighing 400 g. The Kuka robotic arm has various programmable pinching forces (force applied to the object to lift); however, we set it to 40 N to lift the aluminum bar. The pressure sensor attached to the arm as shown in Figure 8a–c represents the aluminum bar being grabbed and released. Corresponding open circuit voltages were measured using a DC source meter, and a graph was plotted as depicted in Figure 9. The sensor generated +13 V and −13 V output voltages during a grab and release operation, and the output voltage response indicates the steady stage operation in the experiment. From graph 9, we can conclude that the pressure sensor has a young modulus of 47.7 kPa and will generate a maximum of 13 V. Conversely, this result may be varied if the device's young modulus has altered or when a greater pinching force is applied. In summary, this study suggests that the pressure sensor built using a low-cost polystyrene polymer material by the electrospinning method can be used as tactile sensors for robots and could replace commercially available expensive pressure sensors.

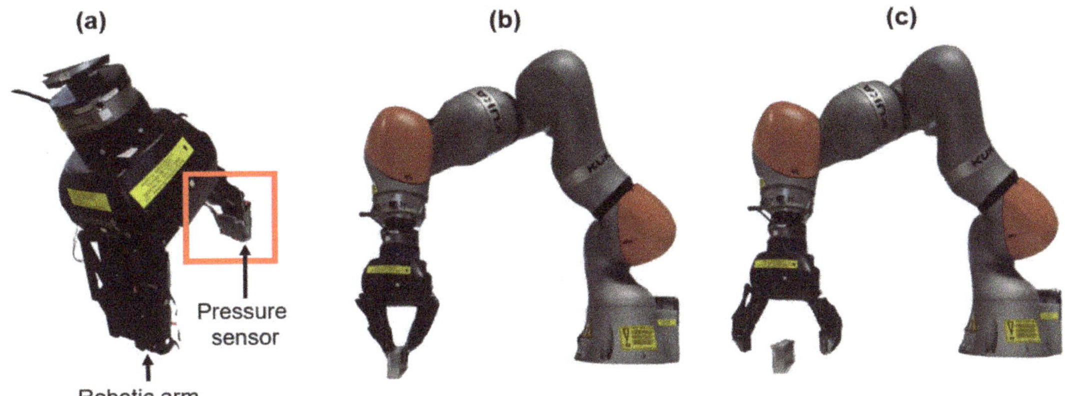

Figure 8. (**a**) Pressure sensor attached to the robotic arm; robotic arm (**b**) grab and(**c**) release mechanism.

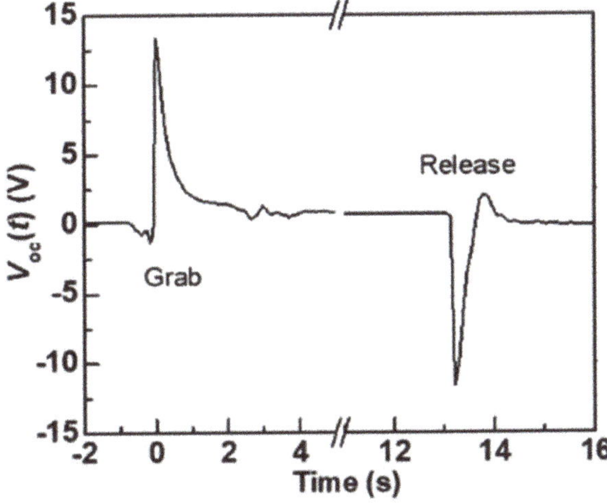

Figure 9. Open circuit voltage with time when the robotic arm grabbed and released the weight.

5. Conclusions

Tactile sensors are vital components to emulate the properties of human skin or electronic skin. Regardless, tactile sensors such as capacitive, optical, magnetic, and piezoelectric types are expensive due to high fabrication costs. On the other hand, converse electromechanical responses were observed inadvertently with electrospun submicron/micron fiber mats composed of nonpiezoelectric polymers, such as poly(d,l-lactic acid (PDLLA), [21] poly(methyl methacrylate) (PMMA) and recently polystyrene (PS). In this article, we demonstrated the piezoelectric effect from atactic polystyrene (aPS) material and fabricated microfibers using an electrospinning method. Surface morphology of microfibers were investigated by FSEM. A pressure sensor was fabricated and the output voltage and current responses were calculated using a DC source meter with respect to different applied pressures. A maximum open circuit voltage of 3.1 V was generated at 19.9 kPa applied pressure and the apparent piezoelectric d constant (d_{app}) was calculated as 12.9 ± 1.8 pC N^{-1}. The mechanical characteristics of the sensor were evaluated and the stress–strain plot is presented. A fabricated pressure sensor was attached to a commercially available robotic arm to mimic the tactile sensing and the corresponding voltage output (13 V) was estimated to lift the 400 g aluminum bar. In summary, combining an electrospinning fabrication method with a low-cost polystyrene polymer material to fabricate a pressure sensor could pave new paths and new research directions and commercial applications.

Author Contributions: Conceptualization, T.S.R., Y.I. and A.C., Fabrication, Y.I., Testing and analysis, T.S.R. and Y.I., writing—original draft preparation, T.S.R. and Y.I., writing—review and editing, Y.I. and A.C., supervision, M.H.A. and S.R., project administration, M.H.A. and S.R., funding acquisition, M.H.A. and S.R. All authors have read and agreed to the published version of the manuscript.

Funding: This research was funded by NUS—Advanced Robotics Centre, and the experiments were carried out at NUS—Centre for Nanofibers and Nanotechnology." and "The APC was funded by the LRF.

Conflicts of Interest: The authors declare no conflict of interest.

References

1. Gafford, J.; Ding, Y.; Harris, A.; McKenna, T.; Polygerinos, P.; Holland, D.; Moser, A.; Walsh, C. Shape deposition manufacturing of a soft, atraumatic, deployable surgical grasper. *J. Med. Devices Trans. ASME* **2014**, *8*, 4–5. [CrossRef]
2. Jentoft, L.P.; Tenzer, Y.; Vogt, D.; Liu, J.; Wood, R.J.; Howe, R.D. Flexible, stretchable tactile arrays from MEMS barometers. In Proceedings of the 2013 16th International Conference on Advanced Robotics (ICAR), Montevideo, Uruguay, 25–29 November 2013. [CrossRef]
3. Tenzer, Y.; Jentoft, L.P.; Howe, R.D. The feel of MEMS barometers: Inexpensive and easily customized tactile array sensors. *IEEE Robot. Autom. Mag.* **2014**, *21*, 89–95. [CrossRef]
4. Wu, Y.; Liu, Y.; Zhou, Y.; Man, Q.; Hu, C.; Asghar, W. A skin-inspired tactile sensor for smart prosthetics. *Sci. Robot.* **2018**, *3*, eaat0429. [CrossRef] [PubMed]
5. Wang, B.; Facchetti, A. Mechanically Flexible Conductors for Stretchable and Wearable E-Skin and E-Textile Devices. *Adv. Mater.* **2019**, *31*, e1901408. [CrossRef] [PubMed]
6. Navaraj, W.; Dahiya, R. Fingerprint-Enhanced Capacitive-Piezoelectric Flexible Sensing Skin to Discriminate Static and Dynamic Tactile Stimuli. *Adv. Intell.Syst.* **2019**, *1*, 1900051. [CrossRef]
7. Lepora, N.F.; Church, A.; Kerckhove, C.; De Hadsell, R.; Lloyd, J. From Pixels to Percepts: Highly Robust Edge Perception and Contour Following Using Deep Learning and an Optical Biomimetic Tactile Sensor. *IEEE Robot. Autom. Lett.* **2019**, *4*, 2101–2107. [CrossRef]
8. Silva, P.; Miguel, P.; Ramos, P.; Postolache, O.; Miguel, J.; Pereira, D. Tactile sensors for robotic applications. *Measurement* **2013**, *46*, 1257–1271. [CrossRef]
9. Chou, H.; Nguyen, A.; Chortos, A.; To, J.W.F.; Lu, C.; Mei, J.; Kurosawa, T.; Bae, W.; Tok, J.B.; Bao, Z. Tactile sensing. *Nat. Commun.* **2015**, *6*, 1–10. [CrossRef]
10. Gao, Y.; Jen, Y.; Chen, R.; Aw, K.; Yamane, D.; Lo, C. Sensors and Actuators A: Physical Five-fold sensitivity enhancement in a capacitive tactile sensor by reducing material and structural rigidity. *Sens. Actuators A Phys.* **2019**, *293*, 167–177. [CrossRef]
11. Yan, X.; Yu, M.; Ramakrishna, S.; Russell, S.J.; Long, Y.Z. Advances in portable electrospinning devices for in situ delivery of personalized wound care. *Nanoscale* **2019**, *11*, 19166–19178. [CrossRef]
12. Phys, J.A.; Yao, K.; Yousry, Y.M.; Wang, J. Open-cell P (VDF-TrFE)/MWCNT nanocomposite foams with local piezoelectric and conductive effects for passive airborne sound absorption Open-cell P (VDF-TrFE)/MWCNT nanocomposite foams with local piezoelectric and conductive effects for passive airbo. *J. Appl. Phys.* **2020**, *127*, 214102. [CrossRef]
13. Ishii, Y.; Yousry, Y.M.; Nobeshima, T.; Iumsrivun, C.; Sakai, H.; Uemura, S.; Ramakrishna, S.; Yao, K. Electromechanically Active As-Electrospun Polystyrene Fiber Mat: Significantly High Quasistatic/Dynamic Electromechanical Response and Theoretical Modeling. *Macromol. Rapid Commun.* **2020**, *41*, 2000218. [CrossRef]
14. Kweon, O.Y.; Lee, S.J.; Oh, J.H. Wearable high-performance pressure sensors based on three-dimensional electrospun conductive nanofibers. *NPG Asia Mater.* **2018**, *10*, 540–551. [CrossRef]
15. Selvan, R.T.; Jayathilaka, W.A.D.M.; Hilaal, A.; Ramakrishna, S. Improved Piezoelectric Performance of Electrospun PVDF Nano‾bers with Conductive Paint Coated Electrode. *Int. J. Nanosci.* **2020**, *19*, 1950008. [CrossRef]
16. Selvan, R.T.; Jia, C.Y.; Jayathilaka, W.A.D.M. Enhanced Piezoelectric Performance of Electrospun PVDF-MWCNT-Cu Nanocomposites for Energy Harvesting Application. *Nano* **2020**, *15*, 2050049. [CrossRef]
17. Ishii, Y.; Kurihara, S. Charge generation from as-electrospun polystyrene fiber mat with uncontacted/contacted electrode. *Appl. Phys. Lett.* **2019**, *115*, 203904. [CrossRef]
18. Xin, Y.; Tian, H.; Guo, C.; Li, X.; Sun, H.; Wang, P.; Lin, J.; Wang, S.; Wang, C. PVDF tactile sensors for detecting contact force and slip: A review. *Ferroelectrics* **2016**, *504*, 31–45. [CrossRef]
19. Tseng, H.; Tian, W.; Wu, W. P(VDF-TrFE) Polymer-Based Thin Films Deposited on Stainless Steel Substrates Treated Using Water Dissociation for Flexible Tactile Sensor Development. *Sensors* **2013**, *13*, 14777–14796. [CrossRef]
20. Nanocomposites, P.B. Modelling and Analysis of Elliptical Cantilever Device Using Flexure Method and Fabrication of Electrospun. *Nano* **2020**, *15*, 2050007. [CrossRef]

21. Nobeshima, T.; Ishii, Y.; Sakai, H.; Uemura, S.; Yoshida, M. Electrospun poly (methyl methacrylate) fibrous mat showing piezoelectric properties. *Jpn. J. Appl. Phys.* **2018**, *57*, 05GC06. [CrossRef]
22. Ishii, Y.; Nobeshima, T.; Sakai, H.; Omori, K.; Uemura, S. Amorphous Electrically Actuating Submicron Fiber Waveguides. *Macromol. Mater. Eng.* **2018**, *303*, 1700302. [CrossRef]
23. Hassan, D.; Ah-Yasari, A.H. Fabrication and studying the dielectric properties of (Polystyrene-copper oxide) nanocomposites for piezoelectric application. *Bull. Electr. Eng. Inform.* **2019**, *8*, 52–57. [CrossRef]
24. Ishii, Y.; Kurihara, S.; Kitayama, R.; Sakai, H.; Nakabayashi, Y.; Nobeshima, T.; Uemura, S. High electromechanical response from bipolarly charged as-electrospun polystyrene fiber mat. *Smart Mater. Struct.* **2019**, *28*, 08LT02. [CrossRef]
25. Araneo, R.; Bini, F.; Rinaldi, A.; Notargiacomo, A.; Pea, M.; Celozzi, S. Thermal-electric model for piezoelectric ZnO nanowires. *Nanotechnology* **2015**, *26*, 265402. [CrossRef]

 *nanomaterials*

Development of an Innovative and Green Method to Obtain Nanoparticles in Aqueous Solution from Carbon-Based Waste Ashes

Raffaella Striani [1], Enrica Stasi [1,2], Antonella Giuri [3], Miriam Seiti [2], Eleonora Ferraris [2] and Carola Esposito Corcione [1,3,*]

1   Dipartimento di Ingegneria dell'Innovazione, Università del Salento, 73100 Lecce, Italy; raffaella.striani@unisalento.it (R.S.); enrica.stasi@unisalento.it (E.S.)
2   Department of Mechanical Engineering, Campus de Nayer, KU Leuven, 2860 Sint-Katelijne-Waver, Belgium; miriam.seiti@kuleuven.be (M.S.); eleonora.ferraris@kuleuven.be (E.F.)
3   Istituto di Nanotecnologia CNR-Nanotec, Polo di Nanotecnologia c/o Campus Ecotekne, Via Monteroni, 73100 Lecce, Italy; antonella.giuri@unisalento.it
*   Correspondence: carola.corcione@unisalento.it

Abstract: In this study, an original and green procedure to produce water-based solutions containing nanometric recycled carbon particles is proposed. The nanometric particles are obtained starting from carbon waste ashes, produced by the wooden biomass pyro-gasification plant CMD (Costruzioni motori diesel) ECO20. The latter is an integrated system combining a downdraft gasifier, a spark-ignition internal combustion engine, an electric generator and syngas cleaning devices, and it can produce electric and thermal power up to 20 kWe and 40 kWth. The carbon-based ashes (CA) produced by the CMD ECO20 plant were, first, characterized by using differential scanning calorimetry (DSC) and microcomputed tomography (microCT). Afterward, they were reduced in powder by using a milling mortar and analyzed by scanning electron microscopy (SEM), energy-dispersive X-ray (EDX) spectrometry, thermogravimetric analysis (TGA), X-ray diffraction (WAXD) and Fourier-transform infrared (FTIR) spectroscopy. The optimization of an original procedure to reduce the dimensions of the ashes in an aqueous solution was then developed by using ball milling and sonication techniques, and the nanometric dimensions of the particles dispersed in water were estimated by dynamic light scattering (DLS) measurements in the order of 300 nm. Finally, possible industrial applications for the nanomaterials obtained from the waste ashes are suggested, including, for example, inks for Aerosol Jet® Printing (AJ® P).

Keywords: recycling; circular economy; nanometric carbon-based ashes; AJ®P

Citation: Striani, R.; Stasi, E.; Giuri, A.; Seiti, M.; Ferraris, E.; Esposito Corcione, C. Development of an Innovative and Green Method to Obtain Nanoparticles in Aqueous Solution from Carbon-Based Waste Ashes. *Nanomaterials* **2021**, *11*, 577. https://doi.org/10.3390/nano11030577

Academic Editor: Ana María Díez-Pascual

Received: 31 January 2021
Accepted: 20 February 2021
Published: 25 February 2021

1. Introduction

During the last decade, the circular economy has become an important domain of academic research, with a sharp increase in the number of articles and journals covering this topic [1]. Geissdoerfer et al. [1], as well as Schut et al. [2], claim that the most significant circular economy definition has been provided by Ellen MacArthur Foundation [3,4]. The use of wastes, including wastes from industrial processes, such as carbon-based ashes (CA), is, therefore, an important development objective of the circular economy [5].

The management of carbon waste ashes still represents an issue. Even though significant amounts of CA are used in various applications, for example, as a substitute for cement in concrete, [6,7], large quantities of them are not utilized and are disposed of in landfills [6], with long-term negative effects on the environment and human health. CA indeed contribute to the formation of particulate matter (PM), which is one of the major causes of airborne pollution and has been found to induce different cancers, cardiovascular diseases and reproductive disorders [8]. Hence, it would become necessary to regard CA as raw material that can be converted into new products rather than waste.

In the last few years, several studies have proposed the reuse of carbon waste ashes to obtain nanomaterials for different applications. As an example, Ramanathan et al. investigated the use of CA, obtained as a byproduct of the coal combustion from power plants, for the synthesis of nanosized particles due to their enrichment in silica, kaolin, iron, and alumina. In this regard, the nanocrystalline aluminosilicates are one of the most promising nanocomposites synthesized starting from CA through the alkaline treatment method. Their main applications are in fields such as wastewater treatment, agriculture system, and antioxidants, although their use is being evaluated in the medical field applications, especially in drug delivery and delivery systems, bone engineering, biosensors, hemodialysis and intestinal therapies [9]. In addition, Salah et al. investigated the production of carbon nanotubes (CNTs) using ultrasonicated CA, obtained by the heavy oil combustion from water desalination plants, as precursors and catalysts [10,11]. These authors also reported on the preparation of carbon nanoparticles (CNPs) using the ball milling technique [12]. These CNPs were useful for various applications, including fillers in epoxy composites, additives in lubricant oil, gas adsorption, etc. [12,13]. Schlatter et al. also investigated the use of carbon black, derived from the combustion of heavy petroleum products, in inkjet printing for the production of electronic devices [14]. However, to the best author's knowledge, no evidence has been reported so far regarding the reuse of CA, derived by power plants for domestic and public use, and/or its use for the development of 3D printing materials.

In this study, an original and green procedure to produce water-based solutions containing nanometric recycled carbon particles is proposed. CA, used in this work, is produced by the wooden biomass pyro-gasification plant CMD ECO20. These plants and processes are applied in the production of electric power and heat for domestic and small public facilities. Overall CA production can be estimated in ~1 kg/h (5 wt % of biomass introduced into the plant) [15]. Similar to coal ashes reported in [9–13], biomass ashes are a multi-component system of powder material [16]. Nevertheless, they show contents of Ag, Au, B, Be, Cd, Cr, Cu, Mn, Ni, Rb, Se and Zn, which are higher than the respective Clarke values (worldwide average contents) for coal ashes [16], making them particularly attractive for applications, such as, production of construction materials and sorbents, but also synthesis and production of minerals, ceramics and others. Hence, the CMD CA were, first, characterized by means of several techniques, including microcomputed tomography (microCT), differential scanning calorimetry (DSC), scanning electron microscopy (SEM), energy-dispersive X-ray (EDX) spectrometry, thermogravimetric analysis (TGA), wide-angle X-ray diffraction (WAXD) and Fourier-transform infrared (FTIR) spectroscopy, in order to identify their composition, size, morphology and material properties. The ashes are then treated by various techniques in order to reach a nanometric size, as requested by the typical applications. The dispersion obtained is finally characterized by dynamic light scattering (DLS). In conclusion, a possible application for the obtained carbon-based water solution as an ink component for Aerosol Jet® Printing (AJ®P) is proposed. AJ®P is a nozzle-based additive manufacturing technology of the direct writing family, aiming at the production of fine features on a wide range of substrates. Originally developed for the manufacture of electronic circuits, AJ®P has been investigated for a range of applications, including active and passive electronic components, actuators, sensors, as well as a variety of selective chemical and biological responses [17]. The use of water-based solutions containing carbon nanomaterials, such as carbon nanotubes [18], [19,20], graphene oxide [21] and graphene [22], for AJ®P is known in the literature. However, to the best of the author's knowledge, recycled carbon nanoparticles aqueous inks have never been using before in AJ®P.

2. Materials and Methods

2.1. Generation of the Byproduct (CA)

Carbon-based ashes (CA) are the byproduct of the biomass pyro-gasification plant CMD ECO20, developed by the company Costruzioni Motori Diesel S.p.A. (CMD) to produce electric power and heat, starting from woodchips waste. The CMD ECO20 plant is an integrated system combining a downdraft gasifier, a spark-ignition internal combustion engine (ICE), an electric generator and syngas cleaning devices. This system processes wooden biomass of G30 size (1.50 to 3.00 cm) and max at 20% of humidity. It can produce electricity and heat up to 20 kWe and 40 kWth, and it is a computerized machine managed at every level of operation [23] and used in domestic and consumer applications. CA are collected and recovered from the dust collector downstream of the plant.

2.2. Characterization of the Byproduct (CA)

The properties of the CA were characterized under various aspects. The ashes collected from the downstream of the plant will be referred to as neat CA. The thermal conductivity of the neat CA was calculated by means of a differential scanning calorimetry (DSC) (Mettler Toledo, Columbus, OH, USA) by using a sensor material (indium), whose melting temperature was 156.6 °C. A neat CA sample was placed into an aluminum crucible, and the indium sensor was put on up the sample. A single scan from 25 °C to 250 °C at a heating rate of 10 °C/min in a nitrogen atmosphere was performed on at least three samples. By considering the method of Flynn and Levin [24], the slopes of the indium and sample endothermic peaks were calculated in order to determine the resistance of the sample (Rs) as follows:

$$Rs = R' - R \tag{1}$$

where R is the thermal resistance between calorimeter and indium, R' is the thermal resistance between calorimeter and indium with the sample. The thermal conductivity (k) is determined by Equation (2):

$$k = \frac{L}{A\,(R' - R)} = \frac{L}{A\,R_s} \tag{2}$$

where L is the sample height, A is the contact area between sample and sensor material.

The porosity of the neat CA was estimated by means of a pycnometer according to Equation (3):

$$\rho = \frac{m}{V} \tag{3}$$

where the m is the mass of the sample and V is the volume calculated according to Equation (4):

$$V = \frac{(m_w + m_s) - m_{ws}}{\rho_w} \tag{4}$$

where m_w is the mass of the water inside the pycnometer, m_s is the dried weight of the sample, m_{ws} is the mass of the system when the sample is immersed in the water, and ρ_w is the density of water (1 g/cm^3).

The microcomputed tomography (microCT) was performed to evaluate the porosity of the neat CA by using a Bruker SkyScan 1172 (Bruker Corporation, Billerica, MA, USA). The following microCT settings were identified for proper scanning of samples, based on their X-ray attenuation capacity: (a) the X-ray source was powered at 50 kV and 200 µA; (b) a 0.5 mm Al filter was used; (c) for the detection of porosity, the pixel size was set at 10 µm; (d) the exposure time was 600 ms, with a 2 × 2 binning; (e) samples were rotated 360°, with a rotation step of 0.4°. After reconstruction with NRecon software, DataViewer was used to visualize the 3D sections of the samples in the XY, XZ and YZ planes. The reconstructed grayscale images of the samples were then analyzed with CTAn software to quantify the porosity after selection of a volume of interest (a cylindrical Volume of Interest-VOI with 5.3 mm diameter and 1 mm height) and appropriate thresholding.

In order to further characterize the neat carbon-based ashes, they were initially reduced in rough powder by using a milling mortar. The ashes treated by milling mortar will be referred to as mortar-milled CA.

The morphology of the mortar milled CA was investigated by scanning microscopy (SEM) using a Zeiss scanning electron microscope Evo40 (Carl Zeiss Microscopy, LLC, White Plains, NY, USA). The elemental analysis of the particles was performed by energy-dispersive X-ray (EDX) spectroscopy using a Bruker, XFlash detector 5010 (Bruker Corporation, Billerica, MA, USA).

Thermogravimetric analysis (TGA) of the mortar milled CA was carried out using a TGA TA instrument SDT Q600 (TA Instruments, New Castle, DE, USA).to confirm the elemental composition. About 10 mg of powder samples were heated in an alumina holder under a nitrogen atmosphere from 20 to 900 °C at a heating rate of 10 °C/min.

The structure of the mortar milled CA was obtained by wide-angle X-ray diffraction (WAXD) using an automatic Bruker D2 Phaser diffractometer (Bruker Corporation, Billerica, MA, USA), in reflection mode, at 35 KV and 40 mA, using nickel-filtered Cu-K radiation (1.5418 Å).

The chemical nature of the functional groups was studied by Fourier-transform infrared (FTIR) spectroscopy of the mortar milled CA using a BRUKER Vertex70 spectrometer (Bruker Corporation, Billerica, MA, USA) equipped with a deuterated triglycine sulfate (DTGS) detector and a KBr beam splitter, at a resolution of 2.0 cm^{-1}. The frequency scale was calibrated to 0.01 cm^{-1} using a He–Ne laser. In total, 32 scans were signal averaged to reduce the noise, and the spectrum of the ashes was collected using KBr pellets.

2.3. Preparation and Characterization of the Aqueous Solution Containing CA-Based Nanoproducts

After the characterization of the CA as raw material (both before and after the mortar milling treatment), the work aimed at reducing the carbon-based ashes to nanometer size in an aqueous solution and investigating their potential for target applications. For this purpose, multistep reduction size processes were developed, monitoring the size and the distribution of the treated CA ashes by multi-angle laser scattering (MALS) (CILAS 1190 particles size analyzer) (CPS Us, Inc., Madison, WI, USA) and by dynamic light scattering (DLS Zetasizer-Malvern) (Malvern Panalytical Ltd, Malvern, UK). In this work, the reduction processes are divided into two principal methods, labeled hereafter, as method I and method II. The ball milling was carried out in an aluminous porcelain jar (1.5 L), using alumina balls in an ambient atmosphere; the mechanical milling was performed in a horizontal oscillatory mill MMS-Ball Mill (M.M.S.2 S.r.l, Nonantola (Modena), Italy), operating at 40 Hz.

The first method, method I, has involved different steps. First of all, the mortar milled CA ashes were dispersed in water. Second, this water-based suspension was ball milled and centrifugated. Finally, the supernatant of the treated CA/water dispersion was sonicated. Based on the results of method I, a second procedure, method II, was proposed in order to reduce the time of the process, improve the efficiency of the size reduction and increase the yield of the process. Therefore, according to method II, the mortar-milled CA were first, dry ball-milled for 24 h and successively dispersed in water and ball-milled for a strongly reduced time (498 h and 24 h, for method I and method II, respectively). After this, a sonication step of the water-CA dispersion was added before the centrifugation step. The time of each step of both methods was established after several experimental measurements of the CA size, as well as the time necessary to obtain the maximum size reduction of the ashes by using the methodology proposed for each step. In detail, the duration of each processing step was identified based on the minimum time necessary to obtain the maximum reduction in size, where the minimum time is identified when two consecutive measurements gave constant size values. In method II, similar micrometric values have been reached in advance compared to method I, significantly reducing the time of treatment. A schematic list of each step of both methods is reported as follows: According to method I, after mortar milling, the CA were dispersed in water and reduced by means of the following protocol:

A. Wet ball milling for 498 h (CA in water: 50 g/L);
B. Centrifuge at 10,000 rpm for 20 min;
C. Sonication of the supernatant for 5 h (by ultrasonic bath CP104) (CA in water: 3 g/L).

According to method II, after mortar milling, the CA were treated by means of the following protocol:

A. Dry ball milling for 24 h;
B. Wet ball milling for 24 h (CA in water: 50 g/L);
C. Sonication for 32 h (by Diagenode Bioruptor Plus sonication device);
D. Centrifuge at 10,000 rpm for 20 min;
E. Sonication of the supernatant for 5 h (by ultrasonic bath CP104) (CA in water: 4 g/L).

3. Results

The neat CA collected by the CMD plant (inset of Figure 1a) are rough and porous particles of irregular shape. The thermal conductivity was measured by DSC analysis (Figure 1a), and it was equal to 0.292 ± 0.003 W/mK. Those values are in line with the ones reported in the literature [25] related to the thermal conductivity of carbon black (0.2–0.3 W/mK), which is widely used as a thermal insulator. From pycnometer measurements, the density was estimated as 0.3 g/cm^3 via Equation (3), and the rough surface was also evinced by microCT analyses (Figure 1b); finally, the quantitative microCT 3D analysis (Figure 1b) revealed an overall porosity of 51.17%.

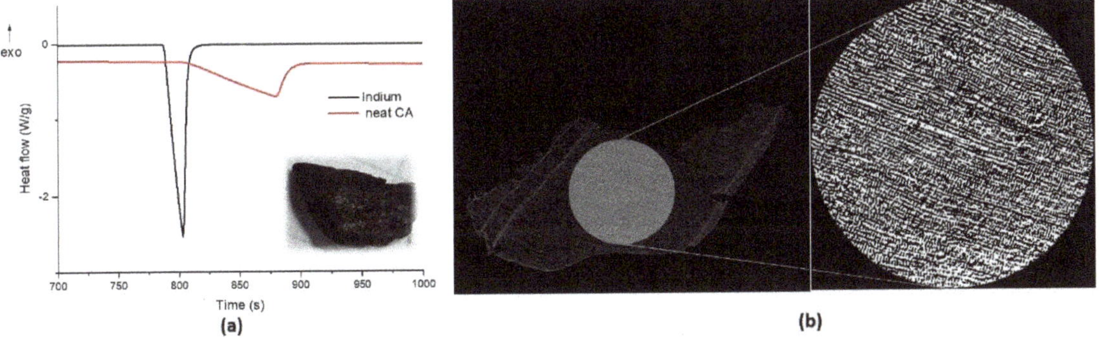

Figure 1. (**a**) Calorimetric curves of thermal conductivity measurements and (**b**) microCT images of neat carbon-based ashes (CA) sample.

Figure 2a–d reports the results on morphological inspection and composition analysis of the milled carbon-based ashes (CA) obtained as a waste product from pyro-gasification of woodchip after initial mortar milling treatment.

The morphological data acquired by scanning electron microscopy (SEM) (Figure 2a) shows particles/fragments with an elongated shape and a minimum average long size of about 80 μm. It is to note that this type of ashes is larger in size than the mortar milled ashes, obtained from previous research conducted by the authors on a different innovative green waste reported in [23], also generated by the CMD plant.

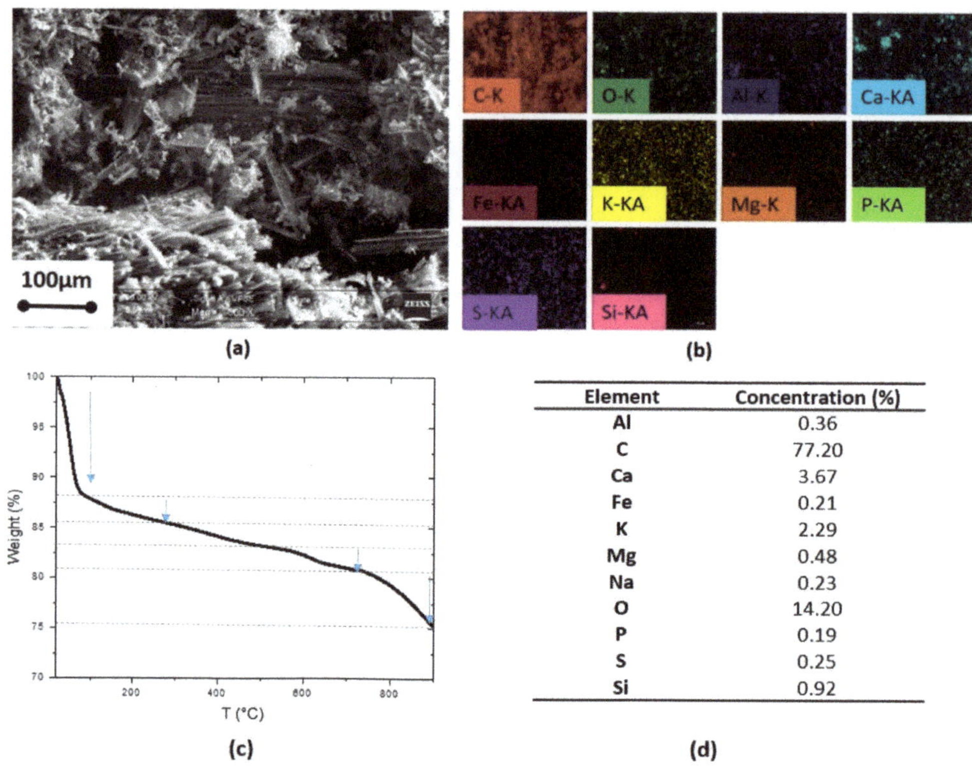

Element	Concentration (%)
Al	0.36
C	77.20
Ca	3.67
Fe	0.21
K	2.29
Mg	0.48
Na	0.23
O	14.20
P	0.19
S	0.25
Si	0.92

(d)

Figure 2. (**a**) SEM image, (**b**) EDX mapping, (**c**) TGA plot whit arrows indicating steps of weight losses and (**d**) EDX data of the carbon-based ashes after mortar milling (mortar milled CA).

The elements analysis performed through the energy-dispersive X-ray (EDX) equipment (Figure 2b) shows that the most abundant element of the CA is carbon, which is present in a percentage of about 77%. Several other atomic elements can also be observed at different concentrations. The most plentiful elements (>1%), besides carbon, are O, Ca and K. Small amounts (<1%) of Al, Fe, Mg, Na, P, S and Si are also present on the ashes surface. The elements characterization evidenced a composition of the ashes from pyro-gasification of woodchips remarkably close to the ashes from pyro-gasification of innovative green waste used in [26] despite the different initial waste products.

The variety of the mortar milled CA composition was further confirmed by thermogravimetric analyses. Indeed, the TGA curve reported in Figure 2c entails several weight loss steps due to the variety of the neat CA composition. In particular, the first step of about 12%, ranging from room temperature to about 100 °C, is attributed to the removal of the molecularly adsorbed water. A second step (about 3%) between 100 °C and 200 °C, is originated from the removal of the thermally labile oxygen-containing functional groups. Another step of about 2% ranging from 580 to about 700 °C, can be due to the decomposition of $CaCO_3$ [27]. A residual weight of about 75% is observed at 900 °C showing that the main component of the ashes is carbon, as also confirmed by the EDX data.

The structural and spectroscopic analysis of mortar milled ashes were made by wide-angle X-ray diffraction (WAXD) and Fourier-transform infrared (FTIR) spectroscopy, as reported in Figure 3a,b, respectively.

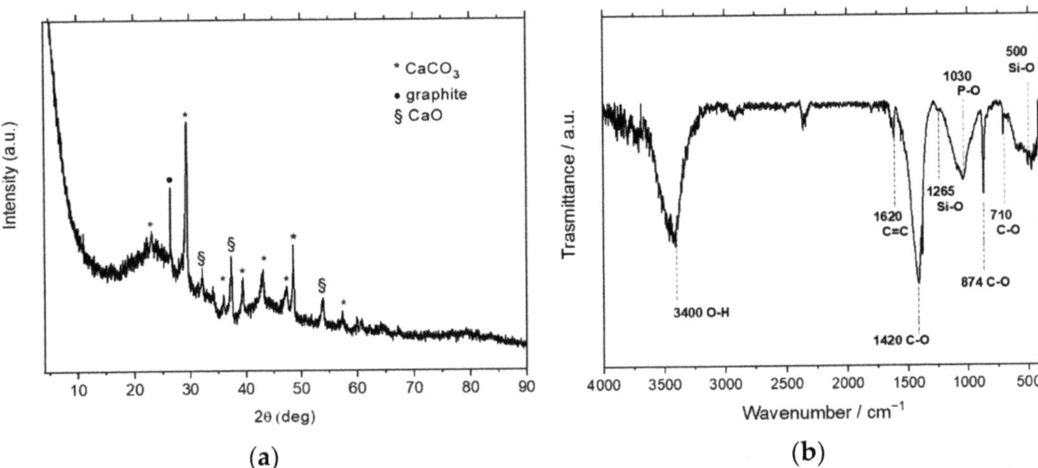

Figure 3. (a) X-ray diffraction and (b) FTIR spectroscopy of the mortar milled CA ashes.

The spectra in Figure 3a show several crystalline peaks, which can be attributed to calcium carbonate ($CaCO_3$) and calcium oxide (CaO), and the typical peak of graphite at $2\theta = 26.5°$ [28]. The diffractometric results are confirmed by FTIR spectra of Figure 3b, where the principal bands related to calcite (1420, 874, 710 cm^{-1}), but also magnesium carbonate (1440, 1375 cm^{-1}), silicates (1260–800 cm^{-1} stretching and 500 cm^{-1} bending) and phosphates (around 1030 cm^{-1}), detected in traces by WAXD, were identified by spectroscopic analysis as reported in [29] for biomass ashes. Moreover, the peak at 3400 cm^{-1} associated with the O-H stretching of the absorbed water confirms the TGA results.

The reduction of particle size of mortar-milled CA for both methods, method I and method II, was evaluated in each phase at regular intervals of time. The results obtained in the first phases of the methods (phase A for method I and phases A, B, C for method II) are reported in Figure 4.

As evidenced by the graphs in Figure 4, the granulometric size of the CA was different for methods I and II. In fact, just after 24 h of dry ball milling (phase A of method II—Figure 4b), the cumulative values at 10%, 50% and 90% of the distributions were lower than the values recorded after 48 h of wet ball milling of method I (Figure 4a). After 48 h (24 h dry ball milling +24 h wet ball milling, i.e., phase A+B in method II), the treated CA particles reached values close to those treated after 186 h by wet ball milling in method I. The same trend was visible throughout the entire curve. In addition, the size of the reduced particles obtained via the steps of Figure 4 are comparable, being the mean diameters equal to (2.03 $\pm$ 0.17) μm in method I and (2.28 $\pm$ 0.08) μm in method II. Nevertheless, method II presents a great advantage in terms of efficiency, reaching comparable resulted to method I in a fraction of time (about 1/6). At this stage, however, none of the two procedures was able to reach a satisfactory quantity of nanometric particles, as from the typical request of industrial applications based on carbon powders. Hence, an additional reducing step was introduced, and the CA treated by methods I and II of Figure 4 were centrifuged at 1000 rpm for 20 min (phase B of method I and phase D of method II, Section 2.3). The results of the granulometric analysis of the supernatants taken after centrifuge are reported in Table 1.

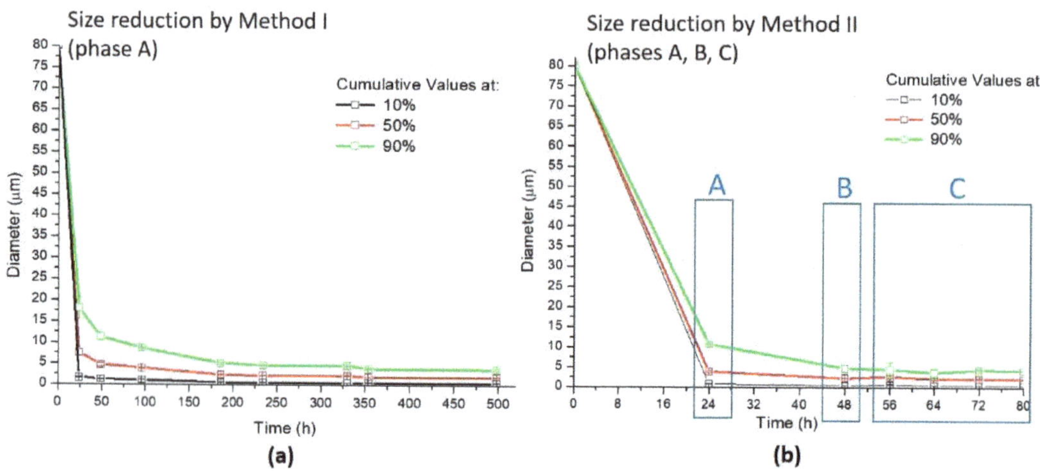

Figure 4. (**a**) Multi-angle laser scattering (MALS) data of the reduced CA during phase A of method I; (**b**) MALS data of the reduced CA during phases A, B and C of method II.

Table 1. Granulometric data of the supernatants were obtained from both methods I and II after centrifuge.

Samples	Cumulative Value at 10% of the Granulometric Distribution (μm)	Cumulative Value at 50% of the Granulometric Distribution (μm)	Cumulative Value at 90% of the Granulometric Distribution (μm)	Mean Diameter (μm)
CA reduced by method I	0.20 ± 0.05	0.96 ± 0.04	1.79 ± 0.09	1.00 ± 0.03
CA reduced by method II	0.10 ± 0.02	0.61 ± 0.28	1.48 ± 0.23	0.69 ± 0.16

Despite the very small size achieved, the majority of the particles were still of micrometer size (90% cumulative value, Table 1). The supernatants were then again treated and subjected to sonication for 5 h in an ultrasonic bath (phases C and E of method I and method II, respectively). It was observed that sonication caused aggregation of the CA particles obtained from method I, changing the esthetical aspect of the solution also at the naked eye (Figure 5a). Instead, the dispersion obtained with treated CA derived from method II appeared stable and homogeneous, revealing particles with a diameter of (331.3 ± 188.7) nm at 98.4% of the distribution, as recorded by means of DLS analysis (Figure 5b). The method II procedure was then revealed to be a simple, ecological, and efficient method able to reduce the carbon ashes from pyro-gasification of wooden biomass to nanometer size particles, dispersed in water in an average concentration of 0.4 wt %. Starting from these preliminary results, method II was selected as the most efficient procedure to obtain aqueous dispersions based on nanometric CA ashes. A representation of the identified optimal procedure (method II) of this paper, step-by-step, along with the granulometric distribution of each phase, is reported in Figure 5b.

Method II, developed in this paper, demonstrated to be able not only to obtain several liters of the water-based dispersions but also to produce dispersions with a reproducible nanometric size. In addition, the nano-based dispersion, obtained by using method II, was also revealed to be stable over time, as demonstrated by the images in Figure 6.

As a provisional study, two inks containing a commercial poly(3,4-ethylenedio-xythiophene)-poly(styrenesulfonate) (PEDOT:PSS) water-based ink (Orgacon® IJ 1005, Agfa, Mortsel, Antwerpen, Belgium and nano-CA water-based dispersion, obtained by the selected method II, were prepared with nano-CA concentrations of 0.05% and 0.1% wt/V$_{PEDOT:PSS}$, respectively. The two inks (each one of ~1 mL) were successfully ultrasoni-cally atomized (Figure 7a) in an AJ®P setup (Optomec, 300s system).

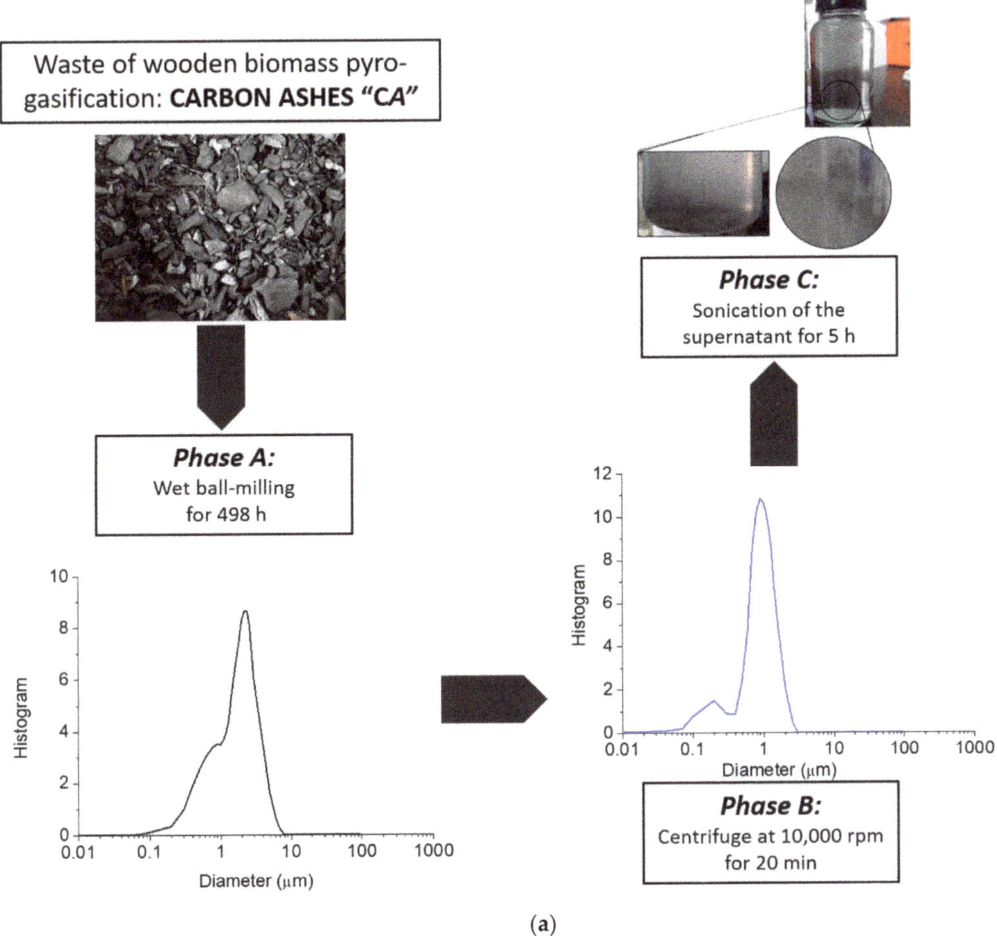

Figure 5. *Cont.*

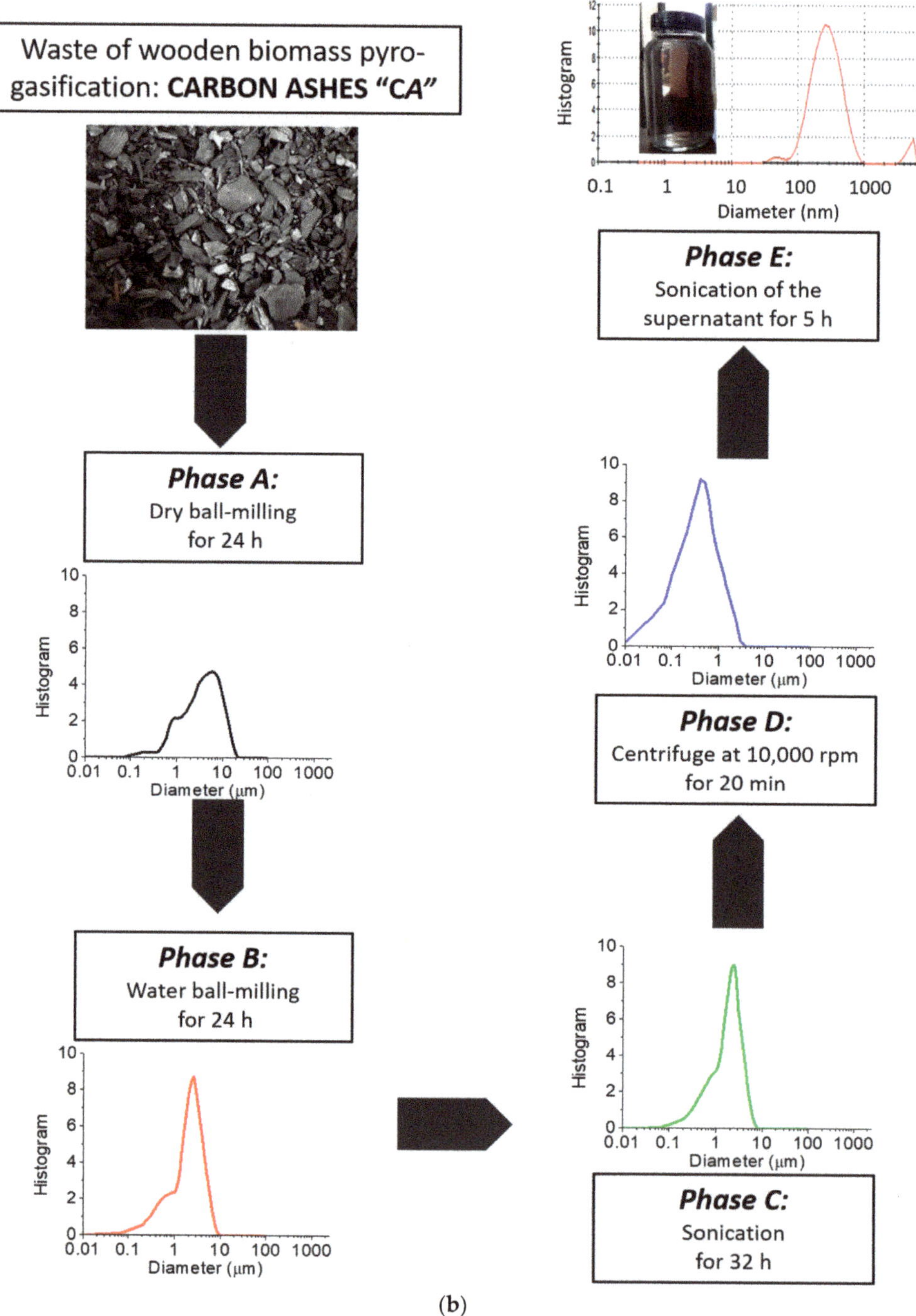

Figure 5. Diagrams of the (**a**) rejected method I and of the (**b**) method II, developed as best practice for the nanometric reduction of carbon ashes generated by pyro-gasification of woodchips and corresponding granulometric analysis.

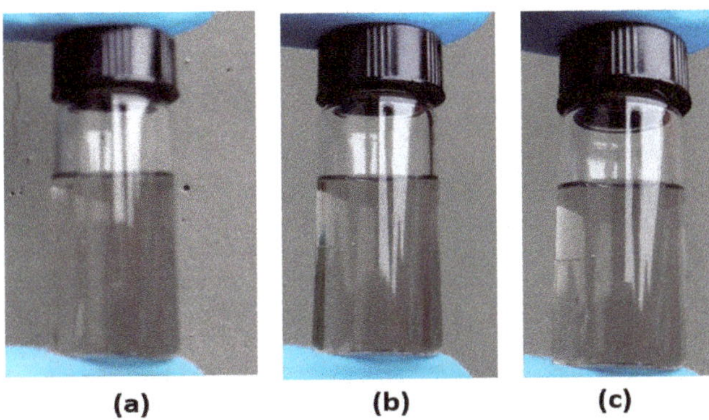

(a)   (b)   (c)

Figure 6. The appearance of Costruzioni Motori Diesel S.p.A. (CMD)-nanoCA-based water dispersion, obtained by method II, after (**a**) 0, (**b**) 30 and (**c**) 90 days.

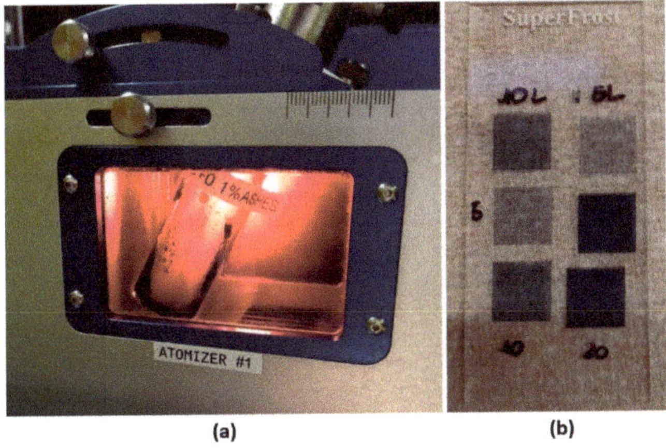

(a)   (b)

Figure 7. (**a**) Ultrasonic atomization in AJ®P setup and (**b**) pattern samples printed onto glass slides of poly(3,4-ethylenedioxythiophene)-poly(styrenesulfonate) (PEDOT:PSS) and nano-CA water-based dispersion with nano-CA concentrations of 0.1% wt/$V_{PEDOT:PSS}$.

4. Discussion

The neat carbon-based ashes (CA) used in this study are produced by the CMD ECO20 plant. This system uses wooden biomass as fuel to provide energy efficiency and environmental benefits by reducing the consumption of fossil fuels and associated greenhouse gas emissions [15]. It finds application in the production of heat and electric power for domestic and public use. The byproduct is a porous, rough carbon ash particle of irregular shape (Figure 1), which was found to have low thermal conductivity (<0.3 W/mK), similar to one of the carbons black, despite the carbon nature. The result is most likely due to the high porosity of the particle collected by the plant. In particular, the density and overall porosity of the neat CA particles of this work were here estimated by pycnometer measurements and microCT analysis, like 0.3 g/cm^3 and 51.17%, respectively. As reported by Wall et al. [30], the thermal conductivity of ash deposits depends in fact not only on the chemical composition, where magnesium or siliceous oxides decrease or increase the thermal conductivity, respectively, but also on the physical state and texture, and the temperature of the deposit. The high porosity of the ashes could be ascribed to the down to few mbar pressures reached during the sintering process. The lack of suitable conditions can, in fact, promote a possible gas production causing the formation of a highly porous structure, as come about to intercalated graphite after the expansion process. Some of the authors studied in previous work [31] how the fast heating at 700 °C for 2 min of the intercalated graphite promoted the formation of a worm-like shape and highly porous microstructure due to the sulfuric acid decomposition and to the formation of intercalated galleries.

The neat CA particles were then pretreated via mortar milling and characterized under various aspects. In particular, the mortar milled ashes composition was found to be very similar to the ones obtained by the CMD plant with different biomass in input, while the size was found to be larger since although both have a lignocellulosic nature, they differ in shape and size, and it is known in the literature that the biomass combustion ashes keep the same morphology of the starting biomass [16], affecting the result obtained by hand milling.

The mortar-milled CA were then further treated to nanometer size, and an original and green procedure to reduce their dimensions was developed (method II) by using various techniques, such as ball milling and sonication, and water-based solutions. Method II was selected among the two procedures explored, as it presents several advantages. The most important advantage is the possibility to obtain stable and reproducible water dispersed nanometric CA-based inks in a shorter time, compared to the time required by method I. This is possible thanks to the initial dry phase, in which the particle size is drastically reduced already within the first 24 h. A further advantage is the possibility to produce carbon-based nanoparticles dispersed in a green solvent (water). In fact, these can be used for many potential industrial applications, or at least for easier, more economic, and more ecological disposal of the nanometric ashes, compared to the neat wastes. Method II was, hence used to produce large amounts (about 5 L) of nanometric water-based dispersions that will be investigated for different potential applications. However, the yield of the method II process is still too low, and this aspect requires further studies. A possible improvement of the proposed methodology could be obtained by using a cryomilling technique by milling the starting mortar milled powders within LN2 with milling balls forming a slurry during milling. This process has been successfully applied in the literature to reduce the dimensions of thermoplastic polymers, and it is known as "cryogenic attrition" [32]. This approach will be tested in a further paper.

Although a low concentration of carbon nanoparticles in aqueous solution (0.4 wt %) could be obtained, by the selected optimal method, method II, these innovative water-based dispersions of nanometric CA can attract considerable attention as material candidates for potential applications in additive manufacturing (AM) or coating technologies and in the sequestration of carbon dioxide by aqueous carbonation [33]. The main features of the proposed CA refer to (i) a particle size around 300 nm for a total of ~98%, (ii) a high presence of carbon (~77%), and (iii) a thermal conductivity like carbon black.

In the context of AM, nanoscale CA dispersions may have a competitive advantage as water-based fillers for the formulation of inks in different printing techniques, such as direct writing (DW) inkjet printing (viscosity range (10–20) mPas, particle size limited by the nozzle diameter, usually $\leq$200 nm), AJ$^{\circledR}$P (viscosity range (1–1000) mPas, particle size $\leq$0.5 µm) and syringe extrusion-based (viscosity range usually (100–10^4) Pas, particle size-dependent by the nozzle, usually $\leq$100 µm) printing [34].

Inks and film coatings, such as organic polymers or ceramics, are indeed widely used in printed electronics (PE) and in the semiconductor industry for conductive circuits (e.g., sensors, thin-film transistors, etc.). Moreover, the embedment of CA nanoparticles into hydrogel-based compositions, such as soft biomaterials or polymers, may increase the mechanical rigidity of the overall structure over time. This property may be useful in biomedical and tissue engineering applications, such as scaffold development for in vitro or in vivo studies.

In this study, two inks, composed of a poly(3,4-ethylenedioxythiophene)-poly(styrene-sulfonate) (PEDOT:PSS) water-based ink (Orgacon$^{\circledR}$ IJ 1005, Agfa, BE) and nanometric CA water-based dispersion with nano-CA concentrations of 0.05% and 0.1% wt/V$_{\mathrm{PEDOT:PSS}}$

were, finally, prepared, in order to verify the potential applications of the nanometric waste ashes. The two CA-based inks were, even, successfully ultrasonically atomized by using an AJ®P setup (Figure 7a). In both cases, homogenous and stable atomization was achieved at power atomization, P = 45 V. Subsequently, pattern samples (squared shape 8 × 8 mm, with serpentine filling) (Figure 7b) were printed onto glass slides (Superfrost®, VWR, Leuven, Belgium at platen temperature 60 °C, printing speed 15 mm/s, carrier and sheath gases flow 30 sccm, and a number of layers equal to 5, 10 and 20. No visual differences in the mist generated were detected over the printing period, demonstrating the stability of the inks. A post-sintering process was implemented in a thermal oven (Heraeus, Hanau, Germany at 150 °C for 45 min. Evaluation of sample properties in terms of print quality, thermal and electrical conductivity is currently ongoing. Therefore, future studies will be towards a more comprehensive evaluation of CA performance towards the target application.

In both cases, excellent adhesion on the desired substrate, high-frequency stability, high permittivity, and good biocompatibility (for biomedical purposes) is required.

In addition, a green application, such as the sequestration of CO_2 in aqueous carbonation, will be investigated. The increasing CO_2 concentration in the Earth's atmosphere is indeed one of the main causes of global warming. Therefore, a technology that could contribute to reducing carbon dioxide emissions is the ex situ mineral sequestration (controlled industrial reactors) of CO_2 [33].

5. Conclusions

In this manuscript, carbon-based ashes (CA) obtained as a waste product from pyrogasification of woodchip were characterized by several techniques and treated to reduction to nanometer size for exploitation in industrial applications. The thermal conductivity measurements of the neat CA showed values similar to carbon black, widely used as a thermal insulator, while the microCT analysis highlighted an overall porosity of 51.17%. The scanning electron microscopy (SEM) of the mortar milled CA showed particles/fragments with an elongated shape and a minimum average long size of about 80 μm. The energy-dispersive X-ray (EDX) spectroscopy also demonstrated that the most abundant element of mortar milled CA is carbon, present in a percentage of about 77%, and the most plentiful elements (>1%), besides carbon, are O, Ca and K. Moreover, the thermogravimetric analysis (TGA) highlighted the presence of several weight losses steps due to the variety of carbon ashes composition: in particular, a residual weight of about 75% at 900 °C, ascribed to the main component of the ashes, which is carbon, as also confirmed by EDX data. Finally, the wide-angle X-ray diffraction (WAXD) revealed several crystalline peaks attributed to calcium carbonate and calcium oxide and the typical peak of graphite. These diffractometric results were then confirmed by FTIR spectra.

Afterward, nanometric CA water-based dispersions were obtained by means of an innovative multistep reduction size process in order to propose a potential industrial application for the nanomaterials, obtained from carbon waste ashes, as inks for additive manufacturing (AM) techniques, such as nozzle-based direct writing (DW) AJ®P.

Author Contributions: Conceptualization, C.E.C.; methodology, C.E.C. and A.G.; software, R.S. and E.S.; validation, C.E.C.; formal analysis, C.E.C. and A.G.; investigation, R.S., E.S. and A.G.; resources, C.E.C. and E.F.; data curation, R.S., E.S. and A.G.; writing—original draft preparation, R.S., E.S., A.G. and M.S.; writing—review and editing, C.E.C. and E.F.; visualization, R.S., E.S. and A.G.; supervision, C.E.C. and E.F.; project administration, C.E.C. and E.F.; funding acquisition, C.E.C. and E.F. All authors have read and agreed to the published version of the manuscript.

Funding: This project was funded by the National Operational Program for Research and Innovation 2014–2020 (CCI 2014IT16M2OP005) of the Ministry of Education, University, and Research.

Acknowledgments: The authors acknowledge with thanks: Ministry of Education, University and Research for the financial support and Domenico Cirillo and Maurizio La Villetta of the company Costruzioni Motori Diesel (CMD) S.p.A. for supplying the carbon ashes; Lorenzo Novembre for its support in carrying out some granulometric measurements and Francesco Montagna for microCT measurements.

Conflicts of Interest: The authors declare no conflict of interest.

References

1. Geissdoerfer, M.; Savaget, P.; Bocken, N.M.P.; Hultink, E.J. The Circular Economy—A new sustainability paradigm? *J. Clean. Prod.* **2017**, *143*, 757–768. [CrossRef]
2. Schut, E.; Crielaard, M.; Mesman, M. Circular economy in the Dutch construction sector. *Rijkswaterstaat Natl. Inst. Public Health Environ.* **2015**, 1–58.
3. Kirchherr, J.; Reike, D.; Hekkert, M. Conceptualizing the circular economy: An analysis of 114 definitions. *Resour. Conserv. Recycl.* **2017**, *127*, 221–232. [CrossRef]
4. The Ellen MacArthur Foundation. Towards a Circular Economy—Economic and Business Rationale for an Accelerated Transition. *Greener Manag. Int.* **2012**. Available online: https://www.ellenmacarthurfoundation.org/publications/towards-a-circular-economy-business-rationale-for-an-accelerated-transition (accessed on 30 January 2021).
5. Li, H.; Bao, W.; Xiu, C.; Zhang, Y.; Xu, H. Energy conservation and circular economy in China's process industries. *Energy* **2010**, *35*, 4273–4281. [CrossRef]
6. Ilic, M.; Cheeseman, C.; Sollars, C.; Knight, J. Mineralogy and microstructure of sintered lignite coal fly ash. *Fuel* **2003**, *82*, 331–336. [CrossRef]
7. Vassilev, S.V.; Baxter, D.; Andersen, L.K.; Vassileva, C.G. An overview of the composition and application of biomass ash. *Fuel* **2013**, *105*, 19–39. [CrossRef]
8. Munawer, M.E. Human health and environmental impacts of coal combustion and post-combustion wastes. *J. Sustain. Min.* **2018**, *17*, 87–96. [CrossRef]
9. Ramanathan, S.; Gopinath, S.C.; Arshad, M.M.; Poopalan, P. Nanostructured aluminosilicate from fly ash: Potential approach in waste utilization for industrial and medical applications. *J. Clean. Prod.* **2020**, *253*, 119923. [CrossRef]
10. Salah, N.A. Method of forming carbon nanotubes from carbon-rich fly ash. U.S. Patent 8,609,189 B2, 17 December 2013.
11. Salah, N.; Al-Ghamdi, A.A.; Memic, A.; Habib, S.S.; Khan, Z.H. Formation of Carbon Nanotubes from Carbon Rich Fly Ash: Growth Parameters and Mechanism. *Mater. Manuf. Process.* **2015**, *31*, 146–156. [CrossRef]
12. Salah, N.; Habib, S.S.; Khan, Z.H.; Alshahrie, A.; Memic, A.; Al-Ghamdi, A.A. Carbon rich fly ash and their nanostructures. *Carbon Lett.* **2016**, *19*, 23–31. [CrossRef]
13. Salah, N.; Alshahrie, A.; Alharbi, N.D.; Abdel-Wahab, M.S.; Khan, Z.H. Nano and micro structures produced from carbon rich fly ash as effective lubricant additives for 150SN base oil. *J. Mater. Res. Technol.* **2019**, *8*, 250–258. [CrossRef]
14. Schlatter, S.; Rosset, S.; Shea, H. Inkjet printing of carbon black electrodes for dielectric elastomer actuators. *Proc. SPIE* **2017**. [CrossRef]
15. Costa, M.; la Villetta, M.; Cirillo, D. Performance Analysis of a Small-Scale Combined Heat and Power System Powered by Woodchips. In Proceedings of the 12th Conference on Sustainable Development of Energy, Water, and Environment Systems, Dubrovnik, Croatia, 4–8 October 2017; pp. 1–12.
16. Vassilev, S.V.; Baxter, D.; Andersen, L.K.; Vassileva, C.G. An overview of the composition and application of biomass ash. Part Phase–mineral and chemical composition and classification. *Fuel* **2013**, *105*, 40–76. [CrossRef]
17. Wilkinson, N.J.; Smith, M.A.A.; Kay, R.W.; Harris, R.A. A review of aerosol jet printing—A non-traditional hybrid process for micro-manufacturing. *Int. J. Adv. Manuf. Technol.* **2019**, *105*, 4599–4619. [CrossRef]
18. Cao, C.; Andrews, J.B.; Franklin, A.D. Completely Printed, Flexible, Stable, and Hysteresis-Free Carbon Nanotube Thin-Film Transistors via Aerosol Jet Printing. *Adv. Electron. Mater.* **2017**, *3*, 1–10. [CrossRef]
19. Jones, C.S.; Lu, X.; Renn, M.; Stroder, M.; Shih, W.-S. Aerosol-jet-printed, high-speed, flexible thin-film transistor made using single-walled carbon nanotube solution. *Microelectron. Eng.* **2010**, *87*, 434–437. [CrossRef]
20. Liu, R.; Ding, H.; Lin, J.; Shen, F.; Cui, Z.; Zhang, T. Fabrication of platinum-decorated single-walled carbon nanotube based hydrogen sensors by aerosol jet printing. *Nanotechnol.* **2012**, *23*, 505301. [CrossRef]
21. Liu, R.; Shen, F.; Ding, H.; Lin, J.; Gu, W.; Cui, Z.; Zhang, T. All-carbon-based field effect transistors fabricated by aerosol jet printing on flexible substrates. *J. Micromech. Microeng.* **2013**, *23*, 65027. [CrossRef]
22. Kwon, Y.-T.; Kim, Y.-S.; Kwon, S.; Mahmood, M.; Lim, H.-R.; Park, S.-W.; Kang, S.-O.; Choi, J.J.; Herbert, R.; Jang, Y.C.; et al. All-printed nanomembrane wireless bioelectronics using a biocompatible solderable graphene for multimodal human-machine interfaces. *Nat. Commun.* **2020**, *11*, 1–11. [CrossRef]
23. Stasi, E.; Giuri, A.; La Villetta, M.; Cirillo, D.; Guerra, G.; Maffezzoli, A.; Ferraris, E.; Corcione, C.E. Development and characterization of innovative carbon-based waste ashes/epoxy composites. *Mater. Today Proc.* **2020**. [CrossRef]
24. Flynn, J.H.; Levin, D.M. A method for the determination of thermal conductivity of sheet materials by differential scanning calorimetry (DSC). *Thermochim. Acta* **1988**, *126*, 93–100. [CrossRef]
25. Khizhnyak, P.E.; Chechetkin, A.V.; Glybin, A.P. Thermal conductivity of carbon black. *J. Eng. Phys. Thermophys.* **1979**, *37*, 1073–1075. [CrossRef]
26. Stasi, E.; Giuri, A.; La Villetta, M.; Cirillo, D.; Guerra, G.; Maffezzoli, A.; Ferraris, E.; Corcione, C.E. Catalytic Activity of Oxidized Carbon Waste Ashes for the Crosslinking of Epoxy Resins. *Polymers* **2019**, *11*, 1011. [CrossRef] [PubMed]

27. Stasi, E.; Giuri, A.; Ferrari, F.; Armenise, V.; Colella, S.; Listorti, A.; Rizzo, A.; Ferraris, E.; Corcione, C.E. Biodegradable Carbon-based Ashes/Maize Starch Composite Films for Agricultural Applications. *Polymers* **2020**, *12*, 524. [CrossRef]
28. Siburian, R.; Sihotang, H.; Raja, S.L.; Supeno, M.; Simanjuntak, C. New Route to Synthesize of Graphene Nano Sheets. *Orient. J. Chem.* **2018**, *34*, 182–187. [CrossRef]
29. Romero, E.; Quirantes, M.; Nogales, R. Characterization of biomass ashes produced at different temperatures from olive-oil-industry and greenhouse vegetable wastes. *Fuel* **2017**, *208*, 1–9. [CrossRef]
30. Parquette, B.; Giri, A.; Brien, D.J.O.; Brennan, S.; Cho, K.; Tzeng, J. *Cryomilling of Thermoplastic Powder for Prepreg Applications*; Army Research Lab Aberdeen Proving Ground Md Weapons And Materials Research Directorate: Aberdeen, MD, USA, 2013.
31. Wall, T.; Bhattacharya, S.; Zhang, D.; Gupta, R.; He, X. The properties and thermal effects of ash deposits in coal-fired furnaces. *Prog. Energy Combust. Sci.* **1993**, *19*, 487–504. [CrossRef]
32. Corcione, C.E.; Freuli, F.; Maffezzoli, A. The aspect ratio of epoxy matrix nanocomposites reinforced with graphene stacks. *Polym. Eng. Sci.* **2012**, *53*, 531–539. [CrossRef]
33. Montes-Hernandez, G.; Pérez-López, R.; Renard, F.; Nieto, J.; Charlet, L. Mineral sequestration of CO2 by aqueous carbonation of coal combustion fly-ash. *J. Hazard. Mater.* **2009**, *161*, 1347–1354. [CrossRef]
34. Salary, R.R.; Lombardi, J.P.; Tootooni, M.S.; Donovan, R.; Rao, P.K.; Borgesen, P.; Poliks, M.D. Computational Fluid Dynamics Modeling and Online Monitoring of Aerosol Jet Printing Process. *J. Manuf. Sci. Eng.* **2016**, *139*, 021015. [CrossRef]

 *nanomaterials*

Article

Atomistic Modelling of Size-Dependent Mechanical Properties and Fracture of Pristine and Defective Cove-Edged Graphene Nanoribbons

Daniela A. Damasceno [1,2], R.K.N.D. Nimal Rajapakse [1,3,*] and Euclides Mesquita [4]

1 School of Engineering Science, Simon Fraser University, Burnaby, BC V5A 1S6, Canada;
   daniela.damasceno@usp.br
2 Department of Materials Physics and Mechanics, Institute of Physics, University of São Paulo,
   Ed. Van de Graaff–ed 10-Grupo SAMPA, Rua do Matão, Travessa R, 187, São Paulo 05508-090, Brazil
3 Department of Civil Engineering, Sri Lanka Institute of Information Technology, Malabe 10115, Sri Lanka
4 Department of Computational Mechanics, and Center for Computational Engineering & Sciences (CCES),
   University of Campinas, Mendeleyev, 200-Cidade Universitária, Campinas, São Paulo 13083-860, Brazil;
   euclides@fem.unicamp.br
* Correspondence: rajapakse@sfu.ca

Received: 31 May 2020; Accepted: 13 July 2020; Published: 21 July 2020

Abstract: Cove-edged graphene nanoribbons (CGNR) are a class of nanoribbons with asymmetric edges composed of alternating hexagons and have remarkable electronic properties. Although CGNRs have attractive size-dependent electronic properties their mechanical properties have not been well understood. In practical applications, the mechanical properties such as tensile strength, ductility and fracture toughness play an important role, especially during device fabrication and operation. This work aims to fill a gap in the understanding of the mechanical behaviour of CGNRs by studying the edge and size effects on the mechanical response by using molecular dynamic simulations. Pristine graphene structures are rarely found in applications. Therefore, this study also examines the effects of topological defects on the mechanical behaviour of CGNR. Ductility and fracture patterns of CGNR with divacancy and topological defects are studied. The results reveal that the CGNR become stronger and slightly more ductile as the width increases in contrast to normal zigzag GNR. Furthermore, the mechanical response of defective CGNRs show complex dependency on the defect configuration and distribution, while the direction of the fracture propagation has a complex dependency on the defect configuration and position. The results also confirm the possibility of topological design of graphene to tailor properties through the manipulation of defect types, orientation, and density and defect networks.

Keywords: cove-edges; defects; fracture; graphene; molecular dynamics; strength

1. Introduction

Graphene nanoribbons [1,2] (GNR) are narrow stripes of graphene with widths normally less than 10 nm. The GNR configuration has two important structural parameters: the width and the edge patterns. The most common edges observed experimentally [2–5] are the armchair and zigzag edges, as shown in Figure 1. GNR with a controllable design of the width and edges can open bandgaps [4], which change its electronic properties [6,7] making them promising for the development of new nanoelectronic devices [8–10]. However, several studies [11] show that the GNR applicability is directly related to its types of edges and width. The control of these structural parameters is a challenging task during the synthesis process. To date, the bottom-up synthesis [7] of GNR is considered the most appropriate approach to produce GNR structures with controllable width and smooth edges [12,13].

This advance in the synthesis process gives GNR the potential to be used in the development of GNR-based nanoelectronics [14].

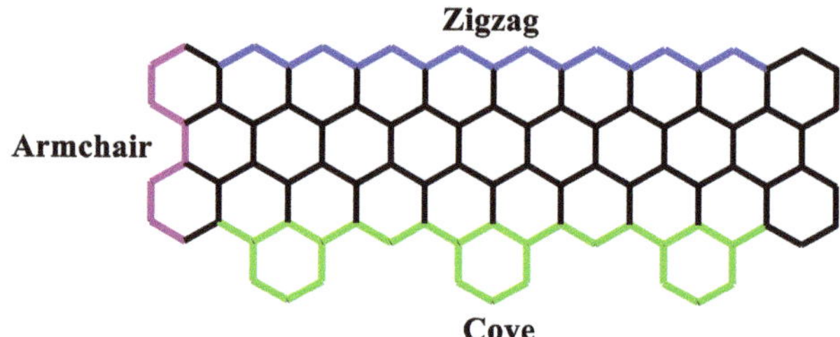

Figure 1. Types of edges in graphene nanoribbon.

Recently, bottom-up synthesis of graphene allowed a configuration of GNR with cove-typed edges [2]. Cove-edged graphene nanoribbons (CGNRs) are a class of nanoribbons with asymmetric edges composed of a repetition of hexagons and vacancies, as shown in green in Figure 1. The cove-typed edges were observed along the zigzag GNR (ZGNR). Since their synthesis, they have been used in several interesting studies involving electronic [15–17] and chemical [18,19] applications.

The armchair (*ar*) and zigzag (*zz*) edges are the most investigated types by both theoretical and experimental approaches. Several studies using molecular dynamics (MD) simulations [20–24], atomic-scale finite element method [25] and experimental measurements [26,27] have been used to investigate the mechanical properties of GNR along the *ar* and *zz* directions. A study of the electronic properties of zigzag graphene nanoribbons with constructed edges representing pentagonal and heptagonal defects has been presented by Pincak et al. [28], whereas the effects of edge vacancies on the electronic properties of zigzag nanoribbons have been addressed by Smotlacha and Pincak [29]. Although the electronic properties of CGNR have been extensively studied [2,15], little attention has been paid to its mechanical properties, which are also very important in device design, fabrication and operation.

Defects cannot be avoided during the synthesis process of GNR; consequently, the applicability of graphene-based materials is also related to the presence of defects. In some cases, defects are introduced into the lattice intentionally, to obtain attractive properties. For example, defects such as Stone–Wales (SW) [30], 5-8-5 [31], extended line of defects (ELD) [32] and nanopores have been introduced in GNR to enhance electronic [27,33–35] and thermal properties [36]. On the other hand, the presence of defects reduces the mechanical strength of GNR [14,31,37], which is prejudicial for the integrity of the structures. Therefore, it is essential to study the effects of defects and edges on the mechanical properties of GNR to support the development of strong GNR-based devices.

Several studies [20,22] using MD simulations have been used to study the mechanical properties and the fracture patterns of the defective *ar* and *oz*. GNR (AGNR and ZGNR, respectively). Results [20] show that topological defects such as Stone–Wales can be introduced into the GNR without compromising the stiffness, whereas the ultimate stress is reduced. Studies [27] also show that defects make GNR more flexible. Moreover, the mechanical properties of GNR with disordered edges [38] and graphene nanowiggles [39] have been studied in details. In regarding to CGNRs, their pristine form showed to possess remarkable electronic properties just by controlling the width and the edges shape. However, studies involving their mechanical properties in the presence of topological and vacancy defects are still missing.

This paper investigates the key mechanical properties of pristine and defective CGNR. It is carried out using MD simulations based on the adaptive intermolecular reactive bond order (AIREBO) potential. Firstly, the mechanical properties of CGNR considering different sizes of width and length are presented. Second, the effects of three different edges configurations on the mechanical response of CGNR are examined. Third, the effect of the presence of topological and vacancy defects on the mechanical response of CGNR is studied. Finally, the effects of the density of divacancy and the combination of vacancy and SW1 defects on the mechanical properties of CGNR are investigated. The present study therefore provides good insight about the mechanical properties of CGNR, which are important in device design and fabrication while enabling device designers to exploit their remarkable electronic properties.

2. Methodology

This study is carried out through MD simulations based on the AIREBO potential [40] using the LAMMPS package [41]. AIREBO potential has been successfully used to model graphene-based nanostructures as well as predict fracture patterns [24,42]. This potential is formed by the sum of three different potentials as presented in Equation (1) [40],

$$E^{AIREBO} = \frac{1}{2} \sum_a \sum_{b \neq a} \left[E_{ab}^{REBO} + E_{ab}^{LJ} + \sum_{c \neq a,b} \sum_{d \neq a,b,c} E_{cabd}^{tors} \right] \tag{1}$$

where the indices a, b, c and d refer to individual atoms; E_{ab}^{REBO} is the reactive empirical bond order (REBO) potential [43], which represents the energy stored in the bond between the atoms a and b; E_{ab}^{LJ} is the Lennard-Jones potential, which considers the non-bonded interactions between atoms and E_{cabd}^{tors} includes the energy from torsional interactions between atoms. In addition, E^{AIREBO} requires a value for the cut-off radii, which is set to 2.0 Å based on the previous studies [24,42].

Figure 2 shows a schematic of cove-edged graphene nanoribbon. The parameter NW corresponds to the number of carbon–carbon dimer lines along the ribbon width (y direction), while the parameter NL corresponds to the number of carbon–carbon dimer lines along the ribbon length (x direction). This convention is in concordance with previous studies [18,44,45]. The NW and NL parameters define the CGNR size. The example in Figure 2 corresponds to NW = 4 and NL = 20. The parameter nH corresponds to the distance between the hexagons at cove-typed edges. For example, 3H means that three hexagons are missing; this parameter is used to represent the cove configuration of the edges. To study the mechanical response of a CGNR, the periodic boundary conditions are considered in the x-direction. Uniaxial tension tests are simulated along the zigzag direction by applying strain at a rate of 0.001 ps^{-1} in the x-direction. All molecular dynamics (MD) simulations are carried out at temperature of 1 K with temperature controlled by the Nose-Hoover thermostat. This thermostat has been widely used for carbon systems and is known to yield stable and efficient computations [20–22,24,37,40,44]. The timestep was 0.5 fs and a uniaxial tension was applied after the system was relaxed at zero pressure through isothermal–isobaric ensemble (NPT).

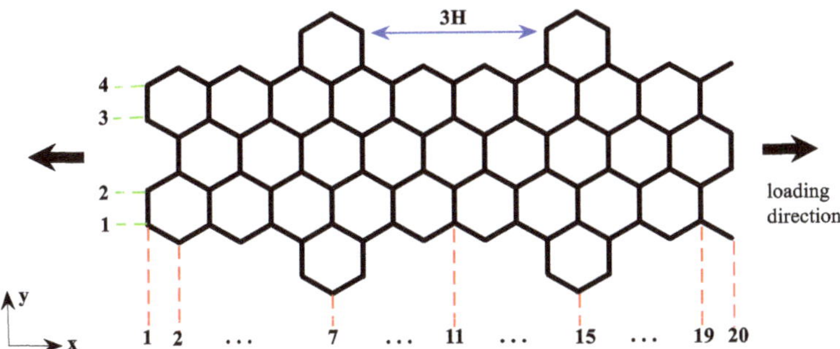

Figure 2. Schematic of cove-edged graphene nanoribbons (CGNR) with the definition of parameters NL (number of carbon–carbon dimer lines along the ribbon length (x direction)), NW (number of carbon–carbon dimer lines along the ribbon width (y direction)) and nH (distance between the hexagons at cove-typed edges).

The results are presented in the form of stress–strain diagrams. The strain measure is defined by the displacement, ΔL_x, divided by the original mesh length L_x^0:

$$\varepsilon_x = \frac{\Delta L_x}{L_x^0} \tag{2}$$

where x represents the Cartesian coordinate axis, $\Delta L_x = L_x - L_x^0$, and L_x is the actual length of the CGNR.

The stress values are obtained by averaging virial stress, σ_{xy} of each carbon atom, expressed by [46]

$$\sigma_{xy} = \frac{1}{V} \sum_{\alpha} \left[\frac{1}{2} \sum_{\beta=1}^{N} \left(R_x^{\beta} - R_x^{\alpha} \right) F_y^{\alpha\beta} - m^{\alpha} v_x^{\alpha} v_y^{\alpha} \right] \tag{3}$$

where the indices x and y denote the Cartesian coordinate axes; V is the volume used for the stress calculation considering the thickness as 3.4 Å; R_x^{β} is the location of atom β along the x axis; R_x^{α} is the location of atom α along the x axis; $F_y^{\alpha\beta}$ is the force acting on atom α due to its interaction with the neighbours β in the y direction; m^{α} is the mass of atom α and v_x^{α} and v_y^{α} are the velocities of atom α along the x and y directions, respectively.

3. Results and Discussion

3.1. Validation of MD Simulation

To validate the numerical model, the stress–strain curves of a pristine bulk graphene sheet under uniaxial tension are compared with the values presented in the literature. A pristine graphene sheet with dimensions of 48.9 Å × 48.4 Å was considered. Figure 3 shows the stress–strain curves along the *ar* and *zz* directions. The ultimate tensile strength obtained from MD simulations was 92 GPa and 115 GPa and the fracture strain was 0.16 and 0.28 along the *ar* and *zz* directions, respectively. The MD results along the *zz* direction agreed well with the experimental [47] results for the fracture stress and fracture strain, which were 130 ± 10 and 0.25, respectively. Moreover, these results were also in agreement with the MD simulation results presented in the literature, ranging from 83 to 137 GPa and 0.12 to 0.27 along the *ar* direction, and 98 to 138 GPa and 0.12 to 0.28 along the *zz* direction [48,49].

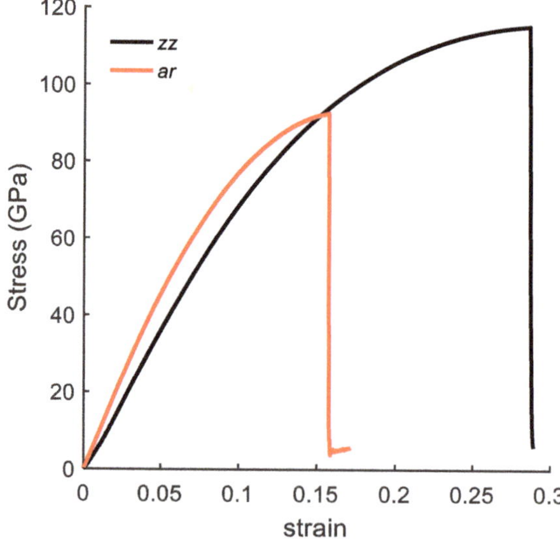

Figure 3. Stress–strain curves of pristine graphene sheet under uniaxial tension along *ar* and *zz* directions.

3.2. Size Effects on Mechanical Response of CGNR

In this section, we focused on the mechanical response of CGNR under uniaxial tension loading considering four different values NL and NW. The CGNR samples have the distance between the hexagons equal to 2H.

Figure 4 presents the size effects on the mechanical response of CGNR in a form of stress–strain curves. Figure 4a shows the stress–strain curves with fixed width of NW = 4, and length varying from NL = 20–68. The mechanical response of CGNR shows hardly any dependency on the length as the periodic boundary conditions were used in the length direction. Such behaviour also confirms the validity and correct physical response of the MD model of CGNR. Moreover, this behaviour was similar to the response previously observed for AGNR and ZGNR [21].

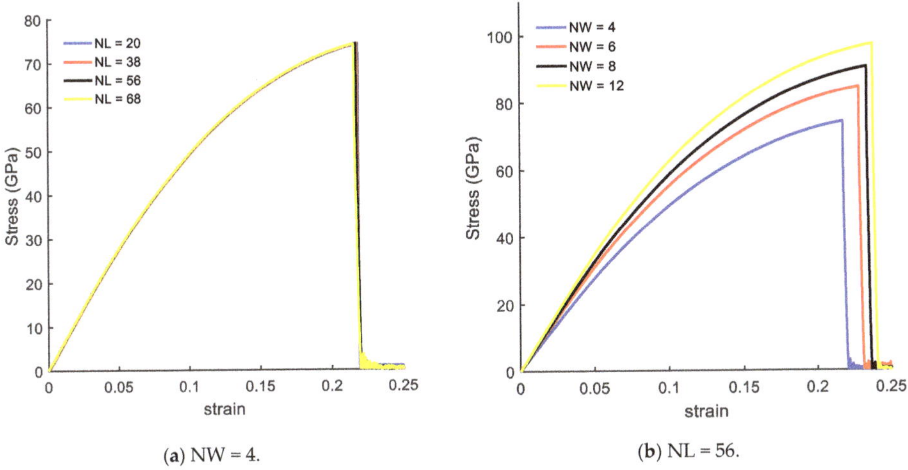

(**a**) NW = 4.

(**b**) NL = 56.

Figure 4. Size effects on the mechanical response of CGNR (2H).

Figure 4b shows the stress–strain curves with a fixed length of NL = 56, and width (NW) varying from 4 to 12. Here, the mechanical response of CGNR shows interesting behaviour with changing widths; CGNR became stronger and more ductile as the width increased. It is due to the edge effects that become significant with decreasing NW and bond breaking becomes easier along the width direction after the first bond breaks. It should be noted that strength and fracture strain of nanoribbons shown in Figure 4 were smaller than the ultimate strength and fracture strain of a pristine sheet shown in Figure 3. However, as NW increased the strain-curve tended to approach the response of a pristine sheet. Comparing pristine zigzag graphene sheet and CGNR with NW = 12 the ultimate tensile strength was 115 GPa and 97 GPa and the failure strain was 0.28 and 0.24, respectively.

Moreover, the mechanical response of CGNR with varying widths was different from the response previously observed for AGNR and ZGNR. Chu et al. [21] and Damasceno et al. [25] concluded that AGNR shows little size dependency with width whereas the size dependency with width is higher in the case of ZGNR. However, both authors [21,25] showed that in the case of the ZGNR as the width increased the ultimate strength and the failure strain decreased. Moreover, other studies also concluded that the ultimate strength is larger as the width is decreased [38,39], while the fracture strain is increased. Therefore, we can conclude that ZGNR with cove-typed edges becomes stronger as the width increases, as well as increase its fracture strain in contrast to the behaviour presented by the ZGNR in the literature. This behaviour is relevant to the design of CGNR devices as the lower mechanical strength and ductility could have a direct effect on the device operation and reliability.

3.3. Edge Effects on the Mechanical Response of CGNR

The results presented in Figure 4 show that the edges played a different role on the mechanical response of CGNR when compared to the effect of edges on the response of AGNR and ZGNR. In this regard, it is important to examine how the behaviour observed in Figure 4 depended on the cove-edge configuration shown in Figure 2 (i.e., the distance between the hexagons). Therefore, in the next example, we investigated the influence of parameter nH by comparing the results for 1H and 3H. For each parameter nH, the CGNR length was fixed with NL = 56 and the width (NW) varies from 4 to 8. These two configurations were called Cove II and Cove III as shown in Figure 5a,b, respectively. In addition, we investigated the influence of higher values of the parameter nH by considering CGNRs with NL = 56 and NW = 4 and NW = 8.

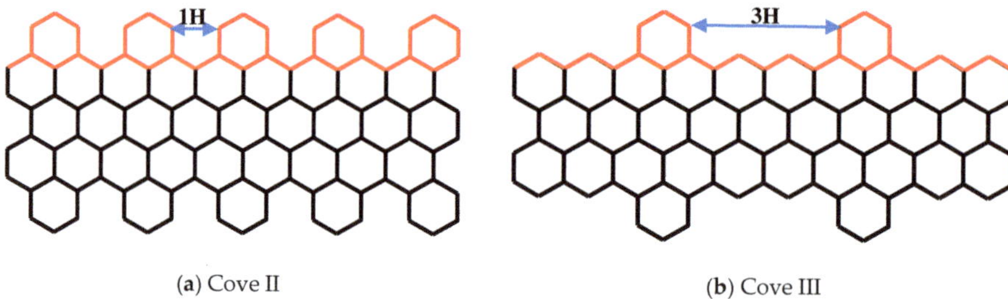

(**a**) Cove II　　　　　　　　　　　　(**b**) Cove III

Figure 5. Schematic of CGNR corresponding to (**a**) 1H and (**b**) 3H.

The mechanical response of the Cove II and Cove III models is presented in Figure 6. Comparing to Figure 4b, the stress–strain curves shown in Figure 6 also demonstrated that as the width increased the CGNR became stiffer irrespective of the nH value but no change in ductility was observed for the lowest value of nH (i.e., Cove II). In the case of Cove III, the stress–strain curves shown in Figure 6b were very close to the curves shown in Figure 4b indicating that the spacing between the hexagons in the cove-edge was not a significant factor in the mechanical response. It is also noted that strength values in Figure 6a were also not substantially different (10%) from the strength values corresponding to 2H and 3H configurations. Moreover, the same behaviour was observed for higher values of nH, as presented in Figure 6c, where dashed lines corresponded to the stress–strain curves of CGNR with width NW = 4 and solid lines with width NW = 8. In both cases, the ultimate fracture stress remained the same for different nH, while the failure strain was slightly increased with increasing n. It could be concluded that, for a given width NW, CGNR presents an almost constant characteristic mechanical behaviour when the distance between the hexagons at the edge were at least spaced by three hexagons, i.e., $n \geq 3$.

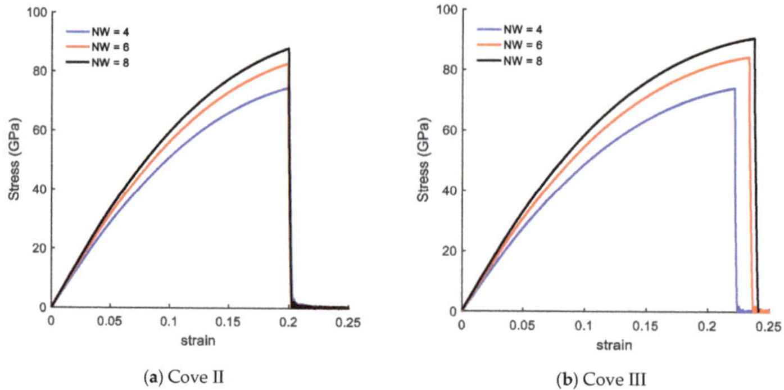

(**a**) Cove II　　　　　　　　　　　　(**b**) Cove III

Figure 6. *Cont.*

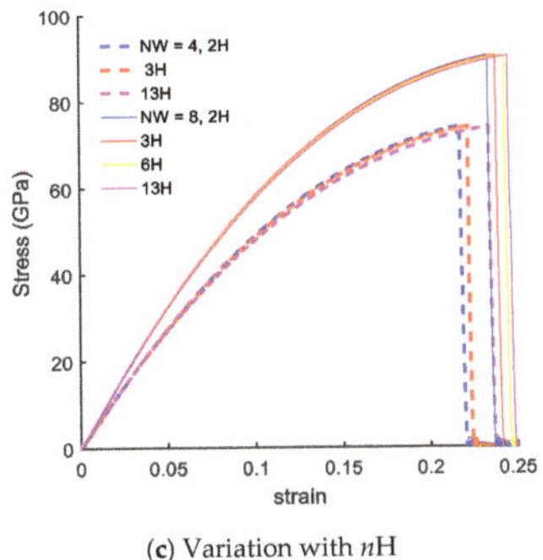

(**c**) Variation with nH

Figure 6. Stress–strain curves of (**a**) Cove II, (**b**) Cove III and (**c**) for different nH values.

To further investigate the effect of the cove-edge geometry, the next example considers a mixed cove-edge configuration with one side identical to the Cove II edge geometry while the opposite edge is represented by the Cove III geometry. The mechanical response of this CGNR is shown in Figure 7. The stress–strain curves shown in Figure 7 show similar behaviour to the curves shown in Figures 4b and 6b; as the width increased the CGNR became stiffer but the variation of ductility with width was negligible. An important observation from Figures 4, 6 and 7 is that the mechanical strength of CGNR showed a minor dependence on the cove-edge configuration with strength and ductility showing less than 15% variation.

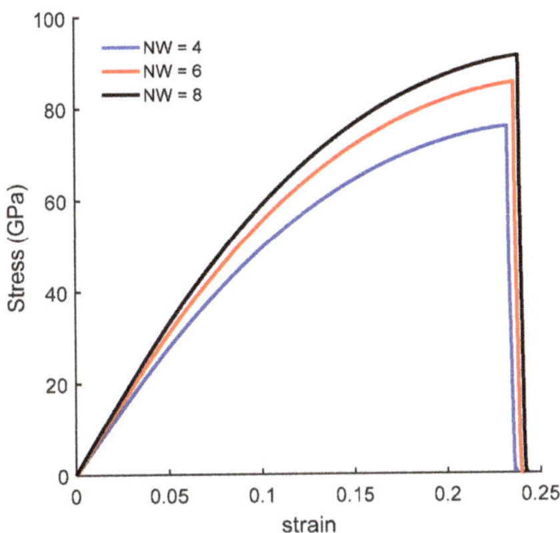

Figure 7. Stress–strain curves of CGNR with mixed edges.

3.4. Topological and Vacancy Defects in CGNR

In this section, the effects of the topological and vacancy defects on the mechanical behaviour of CGNR were presented. The topological defects break the symmetry of the hexagonal lattice, while the sp^2 covalent bonds were maintained. The vacancy defects were obtained by removing atoms from the lattice and as a result the sp^2 covalent bonds around the defects were not maintained. The topological and vacancy defects considered in this study are shown in Figure 8. Figure 8a shows a divacancy (DV) defect obtained by removing two atoms from the center of the lattice. Figure 8b–d shows topological defects known as 5-8-5, 555-777 and 5555-6-7777. Figure 8e,f shows the Stone–Wales defects commonly identified as SW1 and SW2. To study the effect of defects on the mechanical behaviour of CGNR, each defect shown in Figure 8 was placed at the centre of a CGNR with geometry defined by NL = 56 and NW = 8. In all MD simulations, the cove-type edges had hexagons spaced at 2H, and the length of a standard carbon–carbon bond was considered initially as 1.396 Å. Note that initial bond lengths were different for certain bonds and rotated carbon–carbon bonds associated with topological defects.

(a) DV   (b) 5-8-5   (c) 555-777

(d) 5555-6-7777   (e) SW1   (f) SW2

Figure 8. Topological and vacancy defects embedded in CGNR.

Figure 9 shows the mechanical response of CGNR under uniaxial tension loading. The stiffness (elastic modulus) of CGNR shows minor influence of defects and gradually became nonlinear as the strain increased beyond 5%. However, all types of defects had a significant effect on the ultimate strength and ductility of CGNR when compared to a pristine CGNR of identical size. For a pristine CGNR, the ultimate tensile strength was 90.6 GPa and the fracture strain was 0.23. Reductions in strength and fracture strain were highest for the 5-8-5, DV and 555-777 defects followed by the 5555-6-7777, SW2 and SW1. The SW1 configuration was mechanically stronger compared to all other defects. Here, it is interesting to note that the same behaviour was observed for the zigzag graphene with the SW1 defect [50]. Compared to a pristine CGNR, the strength was reduced from 90.6 to 75.7 GPa and the failure strain from 0.23 to 0.15 for a SW1 defect, whereas for a 5-8-5 defect the strength reduced to 61.8 GPa and the strain to 0.12. For a better understanding of the behaviour presented in Figure 9, it is useful to discuss the fracture patterns observed in the Visual Molecular Dynamics package [51] during the MD simulations. Firstly, we discussed the fracture patterns of the defects with higher ultimate stress, which correspond to the defects SW1, SW2 and 5555-6-7777.

Figure 10 shows a pictorial representation of the atomic structure of the CGNR with selected defects obtained from the Visual Molecular Dynamics (VMD) package [51] just prior to the failure point shown in Figure 9. It should be noted that this figure does not show the actual bond lengths but only a visualization of the atoms and bonds that are involved in the failure. Therefore, the red broken lines illustrate the broken bonds of the system, but they do not represent the actual bond lengths. Such representations are commonly used in molecular simulations to illustrate failure patterns, etc., e.g., [22,46]. It is observed that the defects SW1 (Figure 10a) and 5555-6-7777 (Figure 10c) have a horizontal plane of symmetry and two pentagons at the front of the defects. In both cases, only the bonds of four hexagons and two heptagons get stretched and eventually break. However, for a SW1 defect the fracture runs through the two heptagons sparing the pentagons at the fronts as shown in Figure 10d and the bonds shared by the heptagon–heptagon and the heptagon–hexagon were the first to break. For a 5555-6-7777 defect, the fracture does not cut across the defect but involve only the two heptagons on the right side of the defect and the hexagons above and below them as shown in Figure 10f. Here again, the fracture initiates at the bonds shared by heptagon–heptagon and heptagon–hexagon cells. It is interesting to note that the bonds of the central hexagon and none of the pentagons were broken at the failure point. The SW2 defect is like a SW1 defect in terms of the atomic structure of the defective cells but it does not have a horizontal plane of symmetry. As a result, the fracture cuts through five hexagons, two pentagons and one heptagon. The fracture starts on the bonds shared by the pentagon–heptagon and pentagon–hexagon cells. It seems out of the three defects shown in Figure 10, SW1 has the minimum number of defective cells and together with its symmetry make it the strongest and most stretchable configuration followed by SW2 and 5555-6-7777, which has a higher number of defective cells.

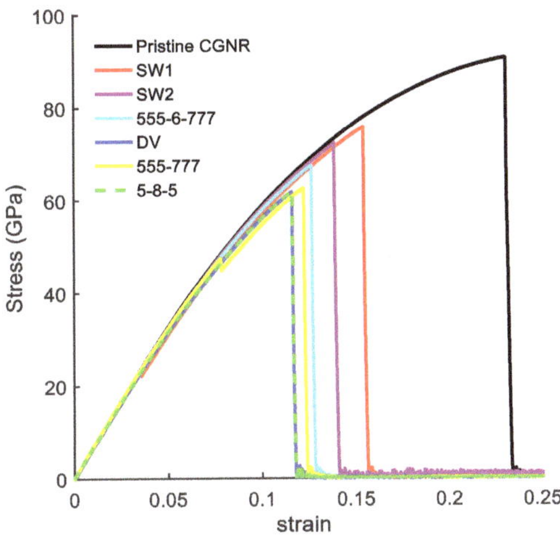

Figure 9. Stress–strain curves for the defects: DV, 5-8-5, 555-777, 5555-6-7777, SW1, and SW2 defects.

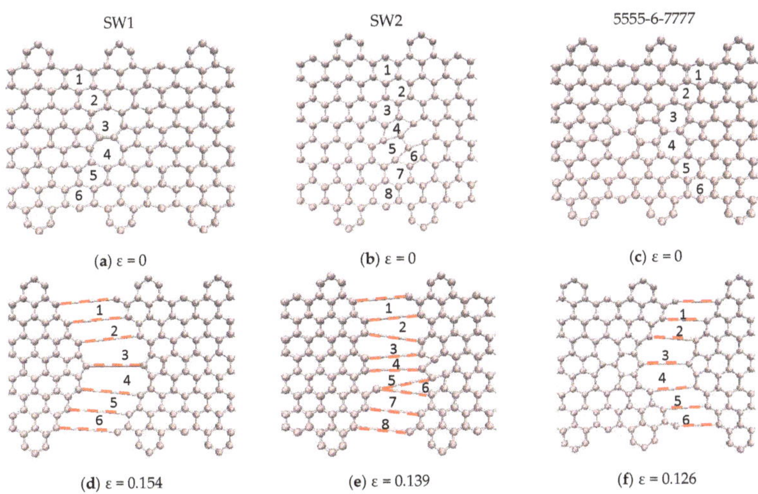

Figure 10. Failure patterns and related atoms and bonds of SW1, SW2 and 555-6-777 defects.

The defects shown in Figure 11a–c had similar fracture propagation as in Figure 10. Here again, the red broken lines illustrate the broken bonds of the system, but they do not represent the actual bond lengths. Fracture propagates through both sides turning to the left side reaching the edges in the interval between of the hexagons. For the 5-8-5 defect, the fracture started on the bonds shared by the pentagon–octagon and the pentagon–hexagon and these bonds aligned with the loading direction relatively easily. In the case of the DV defect, the fracture started on the bonds shared by the vacancy and the hexagons. It obviously had the least resistance to failure as only six bonds were involved, and they aligned easily with the loading direction. The fracture in the 555-777 defect started on the bonds shared by the pentagon–heptagon and pentagon–hexagon. It is interesting to note that in the case of the 555-777 defect, the fracture could start on the bonds shared by the heptagon–heptagon and heptagon–hexagon as in the case of SW1 and 5555-6-7777 defects. However, due to the atomic configuration of the 555-777 defect the fracture started at different bonds. However, its strength and fracture strains were quite close to the values of 5555-6-7777 defect and slightly higher than the DV and 5-8-5 defects.

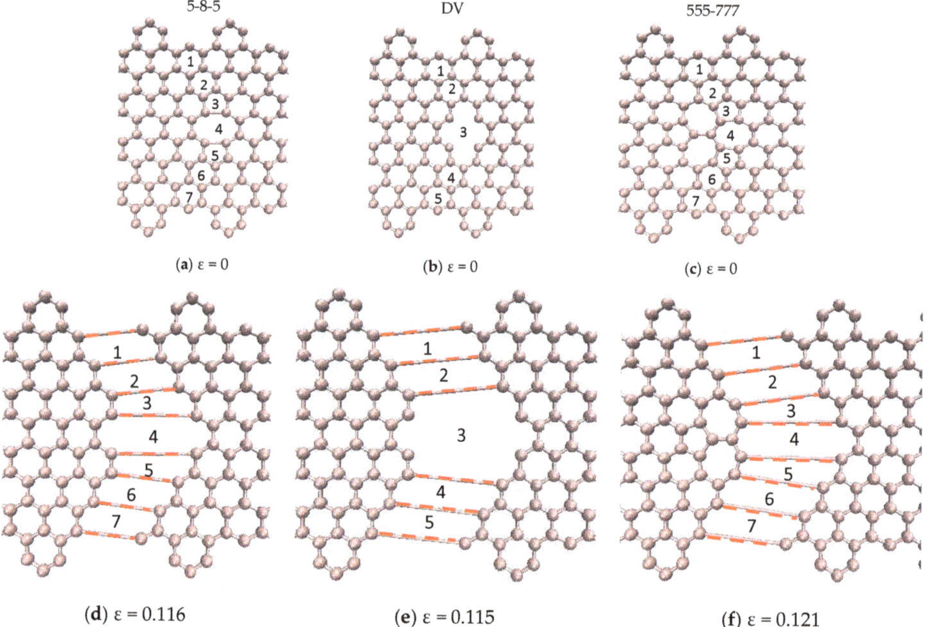

Figure 11. Failure patterns and related atoms and bonds of 5-8-5, DV and 555-777 defects.

Our results revealed that the mechanical response of defective CGNR had a complex dependency on the defect configuration. Highest fracture stress and failure strain were observed for the cases where the fracture started on the bonds shared by heptagon–heptagon and heptagon–hexagon and the number of defective cells was fewer as in the case of SW1, SW2 and 5555-6-7777. Lower fracture stress and failure strain were observed for the cases where the bonds easily aligned parallel to the loading direction during the initial stages of stretching such as in vacancies and pentagon–octagon bonds. Moreover, studies have shown that the bond breaking mechanism is influenced by the amount of covalent bond loading distribution and the corresponding stretching that occurs due to the atom lattice deformation pattern. A discussion on this aspect can be found in Babicheva et al. [52].

3.5. Vacancy Effects on the Mechanical Response of CGNR

One of the most common defects in the GNR [53] is caused by atomic vacancies. Several studies [22,36,54] have examined the effects of vacancies, such as divacancy (DV) and other vacancy configurations, on the mechanical and electronic properties of GNR along the *ar* and *zz* directions. In this section, we extended the single DV defect study presented in the preceding section to examine the mechanical response of a defective CGNR containing a single DV defect with different orientations and multiple DV defects. In addition, we examined the possibility of topological design of CGNR for strength enhancement by combining the DV and SW1 defects, as well as the manipulation of the fracture pattern using these defects. Figure 12 shows six configuration of CGNR with single and multiple DV defects. These configurations are labelled as DV1, DV2, DV3, DV4, DV5 and DV6.

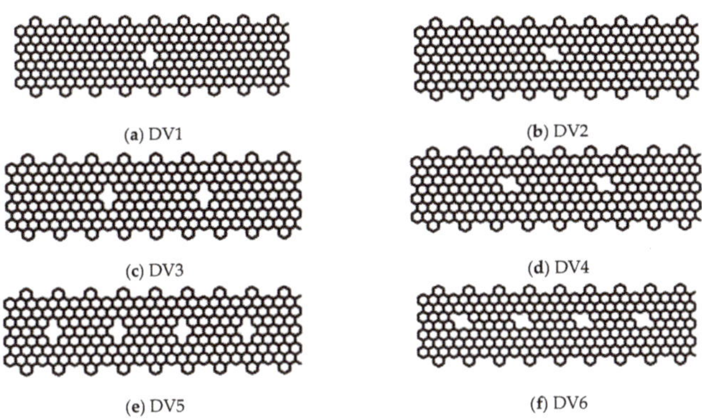

Figure 12. Geometry of CGNR with single and multiple DV defects.

The stress–strain curves of CGNR with single and multiple DV defects, shown in Figure 12, are depicted in Figure 13. Figure 14 compares the fracture patterns of DV 1 and DV 2 defects. It is observed that CGNR with inclined DV defects (DV2, DV4 and DV6) were stronger and more ductile compared to the ribbons with vertical DV defects (DV1, DV3 and DV5). It is also interesting to note that DV 1 had the lowest strength and fracture strain (61.8 GPa and 0.116) followed by DV3 and DV5. In general, the addition of more defects tended to increase the strength slightly and results in a noticeable improvement in ductility. For example, DV6 had the highest strength and ductility (73.9 GPa and 0.165) compared to other five configurations. This observation supported the emerging idea of topological design of 2D materials by manipulating defects to tailor properties.

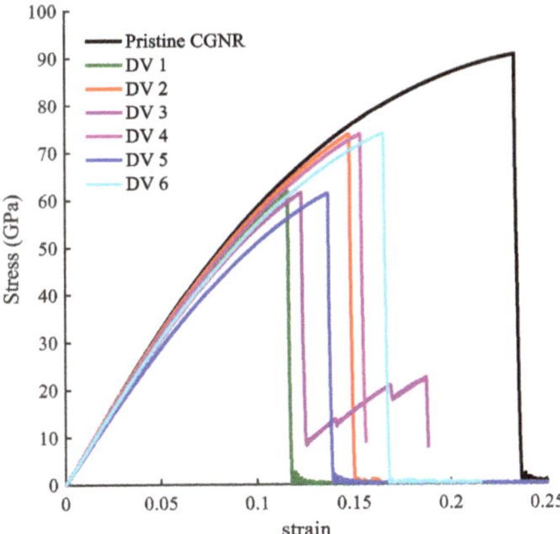

Figure 13. Stress–strain curves of CGNR with single and multiple defects.

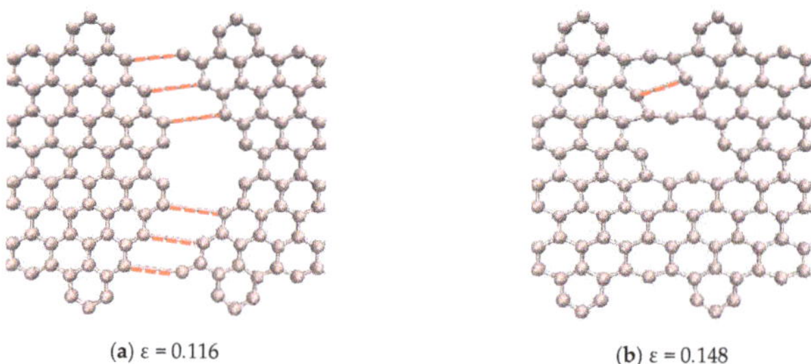

(**a**) ε = 0.116   (**b**) ε = 0.148

Figure 14. Failure patterns and related atoms and bonds of DV1 and DV2 defects.

In the next example, the influence of a network of simple defects was examined to further explore the emerging idea of topological design of 2D materials. We considered a CGNR with a set of inclined DV defects (DV6 configuration in Figure 12 and added SW1 defects to examine the response of the resulting CGNR. Four configurations of CGNR were considered as shown in Figure 15a–d. The mechanical response of these configurations was presented in Figure 15e. The results revealed an interesting behaviour with the combination of the defects. The ultimate strength practically remained the same as the number of SW1 increased, whereas the fracture strain had a noticeable increase. Therefore, it is concluded that the combination of inclined DV and SW1 defects could increase the mechanical flexibility without compromising the stiffness of CGNR.

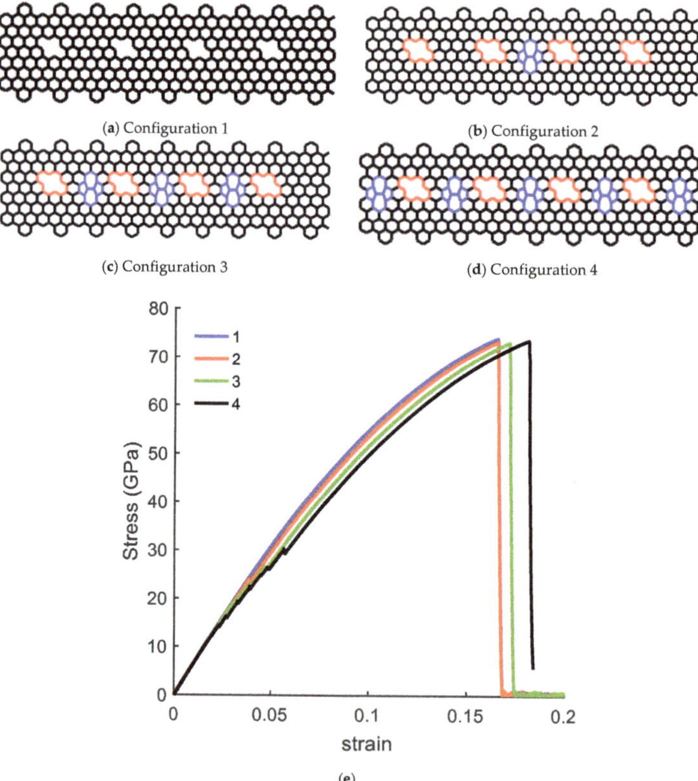

(**a**) Configuration 1   (**b**) Configuration 2

(**c**) Configuration 3   (**d**) Configuration 4

(**e**)

Figure 15. Configuration of CGNR (**a–d**) with defect combinations and (**e**) corresponding stress–strain curves.

4. Conclusions

We investigated the effects of edges, size and defects on the mechanical response of CGNRs by using molecular dynamics simulations. The results revealed that the CGNRs were generally mechanically weaker and less ductile than pristine zigzag graphene. The variation of strength and ductility with ribbon width of CGNR was qualitatively different from a ZGNR. The cove-edge geometry characterized by the distance between the adjacent hexagons shows a relatively minor influence (<15%) on strength and ductility. The mechanical response of defective CGNRs shows complex dependency on the defect configuration, while the fracture pattern was strongly influenced by the defect configuration, position and orientation. The SW1 and SW2 defects were superior to other traditional defects and the DV defects had the lowest strength and fracture strain. It is shown that a change in the orientation of a DV defect could result in a substantial increase in the strength and fracture strain. The presence of multiple DV defects can improve both strength and fracture strain while addition of SW1 defects in between DV defects can improve ductility without compromising strength. The results obtained in this study support the idea of topological design of 2D materials using defects to tailor mechanical properties.

Author Contributions: Conceptualization, methodology, MD implementation, investigation, numerical results, writing—original draft, editing: D.A.D. Methodology, supervision, funding author 1, and manuscript editing: R.K.N.D.N.R.; Conceptualization, supervision, funding author 1, and manuscript editing: E.M. All authors have read and agreed to the published version of the manuscript.

Funding: This study was supported by grants from Natural Sciences and Engineering Research Council of Canada and São Paulo Research Foundation (Fapesp) funding for CEPID Process 2013/08293-7.

Conflicts of Interest: The authors declare no conflict of interest

References

1. Novoselov, K.S.; Geim, A.K.; Morozov, S.V.; Jiang, D.; Zhang, Y.; Dubonos, S.V.; Grigorieva, I.V.; Firsov, A.A. Electric field in atomically thin carbon films. *Science* **2004**, *306*, 666–669. [CrossRef]
2. Liu, J.; Li, B.W.; Tan, Y.Z.; Giannakopoulos, A.; Sanchez-Sanchez, C.; Beljonne, D.; Ruffieux, P.; Fasel, R.; Feng, X.; Müllen, K. Toward Cove-Edged Low Band Gap Graphene Nanoribbons. *J. Am. Chem. Soc.* **2015**, *137*, 6097–6103. [CrossRef]
3. Talirz, L.; Söde, H.; Dumslaff, T.; Wang, S.; Sanchez-Valencia, J.R.; Liu, J.; Shinde, P.; Pignedoli, C.A.; Liang, L.; Meunier, V.; et al. On-Surface Synthesis and characterization of 9-atom wide armchair graphene nanoribbons. *ACS Nano* **2017**, *11*, 1380–1388. [CrossRef]
4. Talirz, L.; Ruffieux, P.; Fasel, R. On-surface synthesis of atomically precise graphene nanoribbons. *Adv. Mater.* **2016**, 6222–6231. [CrossRef]
5. Kobayashi, Y.; Fukui, K.I.; Enoki, T.; Kusakabe, K.; Kaburagi, Y. Observation of zigzag and armchair edges of graphite using scanning tunneling microscopy and spectroscopy. *Phys. Rev. B* **2005**, *71*, 2–5. [CrossRef]
6. Ruffieux, P.; Cai, J.; Plumb, N.C.; Patthey, L.; Prezzi, D.; Ferretti, A.; Molinari, E.; Feng, X.; Müllen, K.; Pignedoli, C.A.; et al. Electronic structure of atomically precise graphene nanoribbons. *ACS Nano* **2012**, *6*, 6930–6935. [CrossRef] [PubMed]
7. Cai, J.; Ruffieux, P.; Jaafar, R.; Bieri, M.; Braun, T.; Blankenburg, S.; Muoth, M.; Seitsonen, A.P.; Saleh, M.; Feng, X.; et al. Atomically precise bottom-up fabrication of graphene nanoribbons. *Nature* **2010**, *466*, 470–473. [CrossRef]
8. Zschieschang, U.; Klauk, H.; Müeller, I.B.; Strudwick, A.J.; Hintermann, T.; Schwab, M.G.; Narita, A.; Feng, X.; Müellen, K.; Weitz, R.T. Electrical characteristics of field-effect transistors based on chemically Synthesized graphene nanoribbons. *Adv. Electron. Mater.* **2015**, *1*, 1–5. [CrossRef]
9. Rizzo, D.J.; Veber, G.; Cao, T.; Bronner, C.; Chen, T.; Zhao, F.; Rodriguez, H.; Louie, S.G.; Crommie, M.F.; Fischer, F.R. Topological band engineering of graphene nanoribbons. *Nature* **2018**, *560*, 204–208. [CrossRef] [PubMed]
10. Puster, M.; Rodríguez-Manzo, J.A.; Balan, A.; Drndić, M. Toward sensitive graphene nanoribbon-nanopore devices by preventing electron beam-induced damage. *ACS Nano* **2013**, *7*, 11283–11289. [CrossRef] [PubMed]
11. Yoneyama, K.; Yamanaka, A.; Okada, S. Mechanical properties of graphene nanoribbons under uniaxial tensile strain. *Jpn. J. Appl. Phys.* **2018**, *57*. [CrossRef]

12. Narita, A.; Feng, X.; Hernandez, Y.; Jensen, S.A.; Bonn, M.; Yang, H.; Verzhbitskiy, I.A.; Casiraghi, C.; Hansen, M.R.; Koch, A.H.R.; et al. Synthesis of structurally well-defined and liquid-phase-processable graphene nanoribbons. *Nat. Chem.* **2014**, *6*, 126–132. [CrossRef] [PubMed]

13. Konnerth, R.; Cervetti, C.; Narita, A.; Feng, X.; Müllen, K.; Hoyer, A.; Burghard, M.; Kern, K.; Dressel, M.; Bogani, L. Tuning the deposition of molecular graphene nanoribbons by surface functionalization. *Nanoscale* **2015**, *7*, 12807–12811. [CrossRef] [PubMed]

14. Vicarelli, L.; Heerema, S.J.; Dekker, C.; Zandbergen, H.W. Controlling defects in graphene for optimizing the electrical properties of graphene nanodevices. *ACS Nano* **2015**, *9*, 3428–3435. [CrossRef] [PubMed]

15. Abbas, A.N.; Liu, G.; Narita, A.; Orosco, M.; Feng, X.; Müllen, K.; Zhou, C. Deposition, characterization, and thin-film-based chemical sensing of ultra-long chemically synthesized graphene nanoribbons. *J. Am. Chem. Soc.* **2014**, *136*, 7555–7558. [CrossRef] [PubMed]

16. Ivanov, I.; Hu, Y.; Osella, S.; Beser, U.; Wang, H.I.; Beljonne, D.; Narita, A.; Müllen, K.; Turchinovich, D.; Bonn, M. Role of edge engineering in photoconductivity of graphene nanoribbons. *J. Am. Chem. Soc.* **2017**, *139*, 7982–7988. [CrossRef] [PubMed]

17. Cao, T.; Zhao, F.; Louie, S.G. Topological phases in graphene nanoribbons: Junction states, spin centers, and quantum spin chains. *Phys. Rev. Lett.* **2017**, *119*, 1–5. [CrossRef]

18. Narita, A.; Chen, Z.; Chen, Q.; Müllen, K. Solution and on-surface synthesis of structurally defined graphene nanoribbons as a new family of semiconductors. *Chem. Sci.* **2019**, *10*, 964–975. [CrossRef]

19. Sisto, T.J.; Zhong, Y.; Zhang, B.; Trinh, M.T.; Miyata, K.; Zhong, X.; Zhu, X.Y.; Steigerwald, M.L.; Ng, F.; Nuckolls, C. Long, atomically precise donor-acceptor cove-edge nanoribbons as electron acceptors. *J. Am. Chem. Soc.* **2017**, *139*, 5648–5651. [CrossRef]

20. Fu, Y.; Ragab, T.; Basaran, C. The effect of Stone-Wales defects on the mechanical behavior of graphene nano-ribbons. *Comput. Mater. Sci.* **2016**, *124*, 142–150. [CrossRef]

21. Chu, Y.; Ragab, T.; Basaran, C. The size effect in mechanical properties of finite-sized graphene nanoribbon. *Comput. Mater. Sci.* **2014**, *81*, 269–274. [CrossRef]

22. Zhang, J.; Ragab, T.; Basaran, C. Comparison of fracture behavior of defective armchair and zigzag graphene nanoribbons. *Int. J. Damage Mech.* **2019**, *28*, 325–345. [CrossRef]

23. Bu, H.; Chen, Y.; Zou, M.; Yi, H.; Bi, K.; Ni, Z. Atomistic simulations of mechanical properties of graphene nanoribbons. *Phys. Lett. A* **2009**, *373*, 3359–3362. [CrossRef]

24. Zhao, H.; Min, K.; Aluru, N.R. Size and chirality dependent elastic properties of graphene Nanoribbons under uniaxial tension. *Nano Lett.* **2009**, *9*, 3012–3015. [CrossRef] [PubMed]

25. Damasceno, D.A.; Mesquita, E.; Rajapakse, R.K.N.D.; Pavanello, R. Atomic-scale finite element modelling of mechanical behaviour of graphene nanoribbons. *Int. J. Mech. Mater. Des.* **2018**, 1–13. [CrossRef]

26. Mazilova, T.I.; Sadanov, E.V.; Mikhailovskij, I.M. Tensile strength of graphene nanoribbons: An experimental approach. *Mater. Lett.* **2019**, *242*, 17–19. [CrossRef]

27. Koch, M.; Li, Z.; Nacci, C.; Kumagai, T.; Franco, I.; Grill, L. How structural defects affect the mechanical and electrical properties of single molecular wires. *Phys. Rev. Lett.* **2018**, *121*, 47701. [CrossRef]

28. Pincak, R.; Smotlacha, J.; Osipov, V.A. Electronic states of zigzag graphene nanoribbons with edges reconstructed with topological defects. *Phys. B Condens. Matter* **2015**, *475*, 61–65. [CrossRef]

29. Smotlacha, J.; Pincak, R. Electronic properties of phosphorene and graphene nanoribbons with edge vacancies in magnetic field. *Phys. Lett. A* **2018**, *382*, 846–854. [CrossRef]

30. Stone, A.J.; Wales, D.J. Theoretical studies of icosahedral C60 and some related species. *Chem. Phys. Lett.* **1986**, *128*, 501–503. [CrossRef]

31. Terrones, H.; Lv, R.; Terrones, M.; Dresselhaus, M.S. The role of defects and doping in 2D graphene sheets and 1D nanoribbons. *Rep. Prog. Phys.* **2012**, *75*. [CrossRef] [PubMed]

32. Lahiri, J.; Lin, Y.; Bozkurt, P.; Oleynik, I.I.; Batzill, M. An extended defect in graphene as a metallic wire. *Nat. Nanotechnol.* **2010**, *5*, 326–329. [CrossRef] [PubMed]

33. Sahan, Z.; Berber, S. Divacancy in graphene nano-ribbons. *Phys. E Low-Dimens. Syst. Nanostruct.* **2019**, *106*, 239–249. [CrossRef]

34. Botello-Méndez, A.R.; Declerck, X.; Terrones, M.; Terrones, H.; Charlier, J.C. One-dimensional extended lines of divacancy defects in graphene. *Nanoscale* **2011**, *3*, 2868–2872. [CrossRef]

35. Tang, G.P.; Zhou, J.C.; Zhang, Z.H.; Deng, X.Q.; Fan, Z.Q. A theoretical investigation on the possible improvement of spin-filter effects by an electric field for a zigzag graphene nanoribbon with a line defect. *Carbon* **2013**, *60*, 94–101. [CrossRef]
36. Kaur, S.; Narang, S.B.; Randhawa, D.K.K. Influence of the pore shape and dimension on the enhancement of thermoelectric performance of graphene nanoribbons. *J. Mater. Res.* **2017**, *32*, 1149–1159. [CrossRef]
37. Zhao, J.; Zeng, H.; Wei, J.; Li, B.; Xu, D. Atomistic simulations of divacancy defects in armchair graphene nanoribbons: Stability, electronic structure, and electron transport properties. *Phys. Lett. A* **2014**, *378*, 416–420. [CrossRef]
38. Tabarraei, A.; Shadalou, S.; Song, J.-H. Mechanical properties of graphene nanoribbons with disordered edges. *Comput. Mater. Sci.* **2015**, *96*, 10–19. [CrossRef]
39. Bizao, R.A.; Botari, T.; Perim, E.; Pugno, N.M.; Galvao, D.S. Mechanical properties and fracture patterns of graphene (graphitic) nanowiggles. *Carbon* **2017**, *119*, 431–437. [CrossRef]
40. Stuart, S.J.; Tutein, A.B.; Harrison, J.A. A reactive potential for hydrocarbons with intermolecular interactions. *J. Chem. Phys.* **2000**, *112*, 6472–6486. [CrossRef]
41. Plimpton, S. Fast parallel algorithms for short-range molecular dynamics. *J. Comput. Phys.* **1995**, *117*, 1–19. [CrossRef]
42. Liu, Y.; Chen, X. Mechanical properties of nanoporous graphene membrane. *J. Appl. Phys.* **2014**, *115*. [CrossRef]
43. Brenner, D.W.; Shenderova, O.A.; Harrison, J.A.; Stuart, S.J.; Ni, B.; Sinnott, S.B. A second-generation reactive empirical bond order (REBO) potential energy expression for hydrocarbons. *J. Phys. Condens. Matter* **2002**, *14*, 783–802. [CrossRef]
44. Osella, S.; Narita, A.; Schwab, M.G.; Hernandez, Y.; Feng, X.; Müllen, K.; Beljonne, D. Graphene nanoribbons as low band gap donor materials for organic photovoltaics: Quantum chemical aided design. *ACS Nano* **2012**, *6*, 5539–5548. [CrossRef] [PubMed]
45. Son, Y.W.; Cohen, M.L.; Louie, S.G. Energy gaps in graphene nanoribbons. *Phys. Rev. Lett.* **2006**, *97*, 1–4. [CrossRef]
46. Dewapriya, M.A.N.; Rajapakse, R.K.N.D.; Phani, A.S. Atomistic and continuum modelling of temperature-dependent fracture of graphene. *Int. J. Fract.* **2014**, *187*, 199–212. [CrossRef]
47. Lee, C.; Wei, X.; Kysar, J.W.; Hone, J. Measurement of the elastic properties and intrinsic strength of monolayer graphene. *Science* **2008**, *321*, 385–388. [CrossRef]
48. Rajasekaran, G.; Narayanan, P.; Parashar, A. Effect of point and line defects on mechanical and thermal properties of graphene: A review. *Crit. Rev. Solid State Mater. Sci.* **2016**, *41*, 47–71. [CrossRef]
49. Cao, G. Atomistic studies of mechanical properties of graphene. *Polymers* **2014**, *6*, 2404–2432. [CrossRef]
50. Damasceno, D.A.; Rajapakse, R.K.N.D.; Mesquita, E.; Pavanello, R. Atomistic simulation of tensile strength properties of graphene with complex vacancy and topological defects. *Acta Mech.* **2020**. [CrossRef]
51. Humphrey, W.; Dalke, A.; Schulten, K. VMD: Visual molecular dynamics. *J. Mol. Graph.* **1996**, *14*, 33–38. [CrossRef]
52. Babicheva, R.I.; Dahanayaka, M.; Liu, B.; Korznikova, E.A.; Dmitriev, S.V.; Wu, M.S.; Zhou, K. Characterization of two carbon allotropes, cyclicgraphene and graphenylene, as semi-permeable materials for membranes. *Mater. Sci. Eng. B* **2020**, *259*, 114569. [CrossRef]
53. Terrones, M.; Botello-Méndez, A.R.; Campos-Delgado, J.; López-Urías, F.; Vega-Cantú, Y.I.; Rodríguez-Macías, F.J.; Elías, A.L.; Muñoz-Sandoval, E.; Cano-Márquez, A.G.; Charlier, J.C.; et al. Graphene and graphite nanoribbons: Morphology, properties, synthesis, defects and applications. *Nano Today* **2010**, *5*, 351–372. [CrossRef]
54. Traversi, F.; Raillon, C.; Benameur, S.M.; Liu, K.; Khlybov, S.; Tosun, M.; Krasnozhon, D.; Kis, A.; Radenovic, A. Detecting the translocation of DNA through a nanopore using graphene nanoribbons. *Nat. Nanotechnol.* **2013**, *8*, 939–945. [CrossRef] [PubMed]

 *nanomaterials*

Article

Multi-Scale Analysis and Testing of Tensile Behavior in Polymers with Randomly Oriented and Agglomerated Cellulose Nanofibers

Fumio Narita *[ID], **Yinli Wang, Hiroki Kurita**[ID] **and Masashi Suzuki**

Department of Materials Processing, Graduate School of Engineering, Tohoku University, Sendai 980-8579, Japan; wang.yinli.r2@dc.tohoku.ac.jp (Y.W.); kurita@material.tohoku.ac.jp (H.K.); masashi.suzuki.t5@dc.tohoku.ac.jp (M.S.)
* Correspondence: narita@material.tohoku.ac.jp

Received: 5 March 2020; Accepted: 2 April 2020; Published: 7 April 2020

Abstract: Cellulose nanofiber (CNF) has been accepted as a valid nanofiller that can improve the mechanical properties of composite materials by mechanical and chemical methods. The purpose of this work is to numerically and experimentally evaluate the mechanical behavior of CNF-reinforced polymer composites under tensile loading. Finite element analysis (FEA) was conducted using a model for the representative volume element of CNF/epoxy composites to determine the effective Young's modulus and the stress state within the composites. The possible random orientation of the CNFs was considered in the finite element model. Tensile tests were also conducted on the CNF/epoxy composites to identify the effect of CNFs on their tensile behavior. The numerical findings were then correlated with the test results. The present randomly oriented CNF/epoxy composite model provides a means for exploring the property interactions across different length scales.

Keywords: multi-scale mechanics; finite element analysis; material testing; cellulose nanofiber; polymer composites; tensile modulus

---

1. Introduction

Cellulose nanofiber (CNF) has been accepted as a valid nanofiller that can improve the mechanical properties of composite materials by both mechanical and chemical methods [1–4]. Nishino et al. [5] determined that the elastic modulus of natural cellulose in the longitudinal direction was 138 GPa. Blefdzki et al. [6] proposed that the ultimate tensile strength of cellulose is estimated to be 17.8 GPa, which is seven times higher than that of steel. Sain and Oksman [7] studied the mechanical properties of CNF to design CNF-reinforced composite materials. Chirayil et al. [8] and Ansari et al. [9] showed that the mechanical properties of CNF-dispersed polymer composites increased compared with those of pure polymer. However, it was reported that an agglomeration of CNF in the matrix, poor adhesion, and water uptake decreased the mechanical properties. Kurita et al. [10] investigated the flexural properties of the CNF-dispersed epoxy layer inserted into woven glass fiber-reinforced polymer (GFRP) composite laminates. They indicated that the flexural strength increases, whereas the flexural modulus is almost constant when inserting two CNF/epoxy interface layers. Xie et al. [11] prepared extremely low CNF-reinforced epoxy matrix composites and tried to understand the strengthening mechanism of the CNF/epoxy composite through a three-point flexural test and finite element analysis (FEA). They proposed the possibility that CNF behaves as an auxiliary agent to enhance the structure of epoxy molecules and not as a reinforcing fiber in the epoxy matrix.

Prediction of the stress and strain requires a detailed FEA. The finite element method is regarded as an important tool in the design and analysis of composite materials, and some finite element works on the modeling of composite materials have been presented by our group [12–14].

In this paper, the tensile behavior of CNF-reinforced polymer composites is investigated. A finite element model that considered the randomly oriented CNF in the polymer was developed, and FEA was performed for the composites with randomly oriented CNFs. Tensile tests were then carried out on the CNF/epoxy composites. The stress–strain curve and Young's modulus for the CNF-reinforced polymer composites were then discussed in detail.

2. Numerical Procedure

2.1. Definition of Representative Volume Element

In this study, ANSYS Multiphysics code (ver. 11.0) was used to perform a three-dimensional FEA. Figure 1 shows the representative volume element (RVE) of the CNF-reinforced polymer composites. As shown in Figure 1a, the RVE was created such that the CNF was embedded in the center of the polymer matrix. The Cartesian coordinate system o-xyz was defined such that the x-axis was in the longitudinal direction, and the y- and z-axes were in the transverse plane of the RVE. L^f and L^m represent the length of the CNF and matrix, respectively. The superscript f represents the fiber, and m denotes the matrix. d^f is the diameter of the CNF. W^m and T^m are the width and thickness of the matrix, respectively. Here, T^m was set to be equal to W^m so that the cross-section yz-plane is square. The CNF aspect ratio L^f/d^f was assumed to be equal to the matrix aspect ratio L^m/W^m [15], and the relationship between W^m and CNF volume fraction V^f was given by the following equation:

$$V^f = \frac{\pi}{4}\left(\frac{d^f}{W^m}\right)^3 \tag{1}$$

The CNF and matrix were assumed to have isotropic elastic properties, and the constitutive equation can be written as

$$
\begin{pmatrix} \varepsilon_{xx}^\delta \\ \varepsilon_{yy}^\delta \\ \varepsilon_{zz}^\delta \\ 2\varepsilon_{yz}^\delta \\ 2\varepsilon_{zx}^\delta \\ 2\varepsilon_{xy}^\delta \end{pmatrix} =
\begin{pmatrix}
1/E^\delta & -v^\delta/E^\delta & -v^\delta/E^\delta & 0 & 0 & 0 \\
-v^\delta/E^\delta & 1/E^\delta & -v^\delta/E^\delta & 0 & 0 & 0 \\
-v^\delta/E^\delta & -v^\delta/E^\delta & 1/E^\delta & 0 & 0 & 0 \\
0 & 0 & 0 & 1/G^\delta & 0 & 0 \\
0 & 0 & 0 & 0 & 1/G^\delta & 0 \\
0 & 0 & 0 & 0 & 0 & 1/G^\delta
\end{pmatrix}
\begin{pmatrix} \sigma_{xx}^\delta \\ \sigma_{yy}^\delta \\ \sigma_{zz}^\delta \\ \sigma_{yz}^\delta \\ \sigma_{zx}^\delta \\ \sigma_{xy}^\delta \end{pmatrix} (\delta = m, f) \tag{2}
$$

where ε_{ij}^δ and σ_{ij}^δ are the strain and stress components ($i, j = x, y, z$), E^δ is the Young's modulus, G^δ is the shear modulus, and v^δ is the Poisson's ratio.

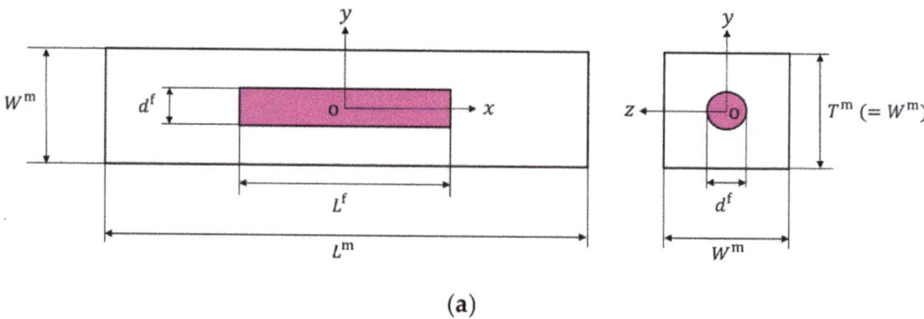

(a)

Figure 1. *Cont.*

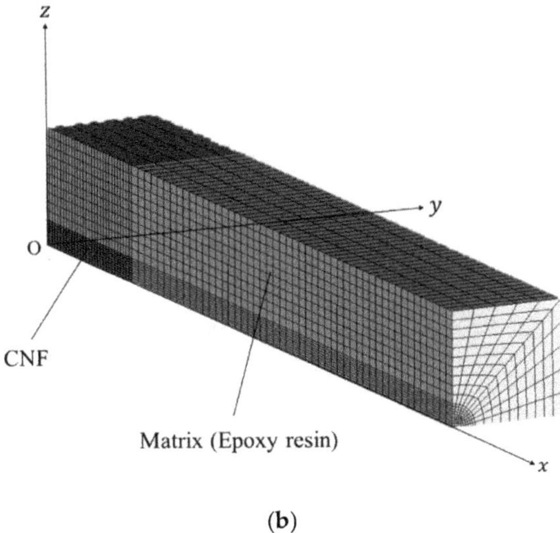

(b)

Figure 1. (**a**) Representative volume element (RVE) of cellulose nanofiber (CNF)-reinforced epoxy composite. (**b**) One-eighth finite element model of RVE for CNF-reinforced epoxy composite.

2.2. Finite Element Analysis of Representative Volume Element

The RVE of CNF-reinforced polymer composites can be assumed to be transversely isotropic with the plane of isotropy being the yz-plane (see Figure 1). The solution obtained from the analysis of the RVE is equivalent to the solution of unidirectional, uniformly dispersed, CNF-reinforced polymer composites. So, for the transversely isotropic RVE, there are five independent elastic properties: longitudinal Young's modulus E_l^r, transverse Young's modulus E_t^r, longitudinal Poisson's ratio v_{lt}^r, transverse Poisson's ratio v_{tt}^r, and longitudinal shear modulus $G_{lt}^r = G_{tl}^r$. The superscript r denotes the RVE. In addition, transverse shear modulus G_{tt}^r is calculated by $G_{tt}^r = E_t^r/2(1 + v_{tt}^r)$. Then, the longitudinal Young's modulus E_l^r and Poisson's ratio v_{lt}^r, transverse Young's modulus E_t^r and Poisson's ratio v_{tt}^r, and longitudinal shear modulus $G_{lt}^r = G_{tl}^r$ can be obtained from the FEA of the RVE under longitudinal normal, transverse normal, and longitudinal shear loadings, respectively.

The appropriate constraints on the RVE under various loads have been determined by Sun and Vaidya [16]. For longitudinal and transverse normal loading, one-eighth of the RVE ($0 \leq x \leq L^m/2$, $0 \leq y \leq W^m/2$, $0 \leq z \leq W^m/2$) was considered, as shown in Figure 1b. The displacement boundary conditions for the RVE under normal load in the longitudinal x-direction are [17]

$$\left.\begin{array}{l} u_x^\delta(0, y, z) = 0 \\ u_x^m(L^m/2, y, z) = u_x^* \end{array}\right\} 0 \leq y \leq W^m/2,\ 0 \leq z \leq W^m/2\ (\delta = m, f) \tag{3}$$

$$\left.\begin{array}{l} u_y^\delta(x, 0, z) = 0 \\ u_y^m(x, W^m/2, z) = u_y^0 \end{array}\right\} \quad 0 \leq x \leq L^m/2, \quad 0 \leq z \leq W^m/2 \quad (\delta = m, f) \tag{4}$$

$$\left.\begin{array}{l} u_z^\delta(x, y, 0) = 0 \\ u_z^m(x, y, W^m/2) = u_z^0 \end{array}\right\} \quad 0 \leq x \leq L^m/2, \quad 0 \leq y \leq W^m/2 \quad (\delta = m, f) \tag{5}$$

where u_x^δ, u_y^δ, and u_z^δ are the displacement components in the x-, y-, and z- directions, respectively; u_x^* is the applied uniform displacement in the x-direction; and u_y^0 and u_z^0 are the uniform displacements in the y- and z-directions. The uniform displacements in the y- and z-directions are determined from the condition that resultant stresses at $y = W^m/2$ plane and $z = W^m/2$ plane are zero, respectively.

The longitudinal Young's modulus $E_l^r = E_x^r$ and Poisson's ratio $v_{lt}^r = v_{xy}^r = v_{xz}^r$ are given by the following equations [16]:

$$E_x^r = \frac{\sigma_{xx}^*}{u_x^*/(L^m/2)} \tag{6}$$

$$v_{xy}^r = -\frac{u_y^0/(W^m/2)}{u_x^*/(L^m/2)} \tag{7}$$

where σ_{xx}^* is the mechanical mean stress on the $x = L^m/2$ plane of the RVE. The mechanical mean stress σ_{xx}^* is obtained as follows [16,17]

$$\sigma_{xx}^* = \frac{\int_0^{W^m/2} \int_0^{W^m/2} \sigma_{xx}^m(L^m/2, y, z)dydz}{(W^m/2)^2} \tag{8}$$

Transverse normal loading can be simulated by applying the uniform displacement in the y- or z-direction. An analysis procedure like that used for the case of longitudinal normal loading can be employed for the case of transverse normal loading, and the transverse Young's modulus $E_t^r = E_y^r = E_z^r$ and Poisson's ratio $v_{tt}^r = v_{yz}^r = v_{zy}^r$ can be obtained.

For longitudinal shear loading, a whole RVE $(-L^m/2 \leq x \leq L^m/2, -W^m/2 \leq y \leq W^m/2, -W^m/2 \leq z \leq W^m/2)$ was considered. The displacement boundary conditions for the RVE under longitudinal shear load are [17]

$$\left.\begin{array}{l} u_x^m(x, y, -W^m/2) = 0 \\ u_y^m(x, y, -W^m/2) = 0 \\ u_z^m(x, y, -W^m/2) = 0 \\ u_x^m(x, y, W^m/2) = u_x^* \\ u_y^m(x, y, W^m/2) = 0 \\ u_z^m(x, y, W^m/2) = 0 \end{array}\right\} \quad -L^m/2 \leq x \leq L^m/2, \quad -W^m/2 \leq y \leq W^m/2 \tag{9}$$

In addition, the points on the $x = -L^m/2$ and $x = L^m/2$ planes with the same y- and z-coordinates need to be constrained to have the same displacements in all three directions as follows [16]:

$$\left.\begin{array}{l} u_x^m(-L^m/2, y, z) = u_x^m(L^m/2, y, z) \\ u_y^m(-L^m/2, y, z) = u_y^m(L^m/2, y, z) \\ u_z^m(-L^m/2, y, z) = u_z^m(L^m/2, y, z) \end{array}\right\} \quad -W^m/2 \leq y \leq W^m/2, \quad -W^m/2 \leq z \leq W^m/2 \tag{10}$$

The longitudinal shear modulus $G_{lt}^r = G_{tl}^r = G_{zx}^r$ is given by [16]

$$G_{zx}^r = \frac{\sigma_{zx}^*}{u_x^*/W^m} \tag{11}$$

The mechanical mean stress σ_{zx}^* is obtained by [16,17]

$$\sigma_{zx}^* = \frac{\int_{-W^m/2}^{W^m/2} \int_{-L^m/2}^{L^m/2} \sigma_{zx}^m(x, y, W^m/2)dxdy}{L^m W^m} \tag{12}$$

In the analysis, the interface between the CNF and matrix was assumed to be perfectly bonding. The eight-node, three-dimensional solid elements were employed to mesh the RVE. The RVE model under longitudinal and transverse normal loadings consisted of 62,736 nodes and 1400 elements, while the model for longitudinal shear loading had 229,869 nodes and 55,680 elements.

2.3. Definition of Randomly Oriented CNF-reinforced Composite

Figure 2a shows the image of the polymer composite with randomly oriented CNF in the FEA. The global Cartesian coordinate system O-*XYZ* and the local Cartesian coordinate systems o-*xyz* for the composite are shown. To create a model of randomly oriented CNF-reinforced polymer composite, the mechanical properties of the RVE denoted by the superscript r were applied to each element, and all local coordinate systems were rotated randomly for the global coordinate system, as shown in Figure 2b, although the geometries of the element and unit cell are different. This random rotation of each element coordinate system was conducted based on the following equation:

$$\begin{pmatrix} x \\ y \\ z \end{pmatrix} = \begin{pmatrix} \cos\varphi & 0 & \sin\varphi \\ 0 & 1 & 0 \\ -\sin\varphi & 0 & \cos\varphi \end{pmatrix} \begin{pmatrix} 1 & 0 & 0 \\ 0 & \cos\omega & -\sin\omega \\ 0 & \sin\omega & \cos\omega \end{pmatrix} \begin{pmatrix} \cos\theta & -\sin\theta & 0 \\ \sin\theta & \cos\theta & 0 \\ 0 & 0 & 1 \end{pmatrix} \begin{pmatrix} x_0 \\ y_0 \\ z_0 \end{pmatrix} \tag{13}$$

where (x_0, y_0, z_0) represents the initial axes before rotation. The initial orientation of the local coordinate systems was the same as that of the global coordinate system, and ω, φ, and θ are the Euler angles indicating the orientation of the local coordinates (x, y, z) with respect to the global coordinates (X, Y, Z). Furthermore, these were generated from random numbers for each element. By using the above definition, it could be possible to simulate randomly oriented CNF-reinforced polymer composites.

The polymer composite with randomly oriented transversely isotropic RVEs (randomly oriented CNFs) is isotropic material [18]. Hence, the constitutive equation for the composite can be written in the following form:

$$\begin{pmatrix} \varepsilon_{XX}^c \\ \varepsilon_{YY}^c \\ \varepsilon_{ZZ}^c \\ 2\varepsilon_{YZ}^c \\ 2\varepsilon_{ZX}^c \\ 2\varepsilon_{XY}^c \end{pmatrix} = \begin{pmatrix} 1/E^c & -v^c/E^c & -v^c/E^c & 0 & 0 & 0 \\ -v^c/E^c & 1/E^c & -v^c/E^c & 0 & 0 & 0 \\ -v^c/E^c & -v^c/E^c & 1/E^c & 0 & 0 & 0 \\ 0 & 0 & 0 & 1/G^c & 0 & 0 \\ 0 & 0 & 0 & 0 & 1/G^c & 0 \\ 0 & 0 & 0 & 0 & 0 & 1/G^c \end{pmatrix} \begin{pmatrix} \sigma_{XX}^c \\ \sigma_{YY}^c \\ \sigma_{ZZ}^c \\ \sigma_{YZ}^c \\ \sigma_{ZX}^c \\ \sigma_{XY}^c \end{pmatrix} \tag{14}$$

where the superscript c represents the composite. The mechanical properties of the composites denoted by the superscript c were obtained using the mechanical properties of the RVE denoted by the superscript r.

2.4. Finite Element Analysis of Randomly Oriented CNF-reinforced Composite Specimen

We performed the FEA by using a one-eighth model of a JIS K 7164 tensile specimen, as shown in Figure 3. The global Cartesian coordinate system O-*XYZ* was employed, and the local coordinate system o-*xyz* was defined (not shown here) and rotated randomly. $L^c = 215$ mm represents the total length; $L_1^c = 60$ mm and $L_2^c = 44.3$ mm are the lengths of the reduced section and the grip section, respectively; $W_1^c = 45$ mm and $W_2^c = 25$ mm are the widths of the reduced section and the grip section, respectively; $T^c = 2$ mm is the total thickness; and $r^c = 60$ mm represents the curvature radius.

The boundary conditions are represented by the following equations:

$$u_X^c(0, Y, Z) = 0 \tag{15}$$

$$u_Z^c(X, Y, 0) = 0 \tag{16}$$

$$u_X^c(L^c/2, Y, Z) = u_X^{**} \tag{17}$$

where u_X^c, u_Y^c, u_Z^c represent the displacement components in X-, Y-, and Z- directions, and u_X^{**} denotes the applied displacement.

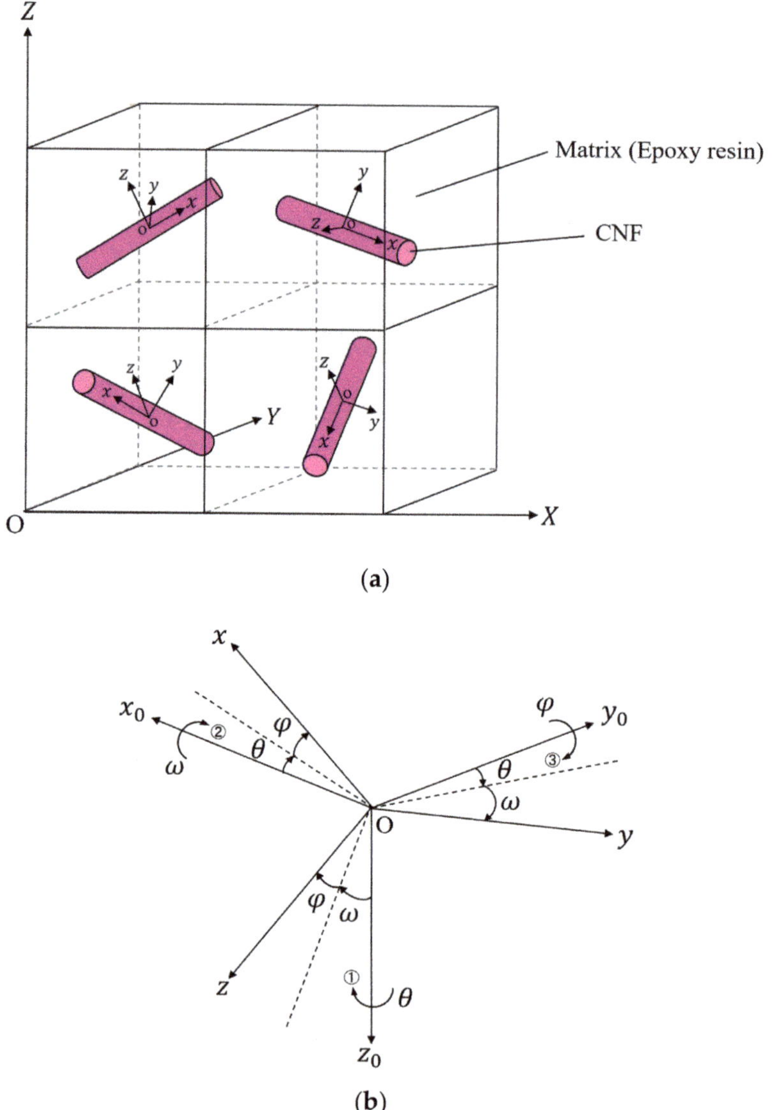

Figure 2. Image of (a) the polymer composite with randomly oriented CNF in the FEA and (b) the rotation of coordinates.

For the finite element model, the eight-node, three-dimensional solid elements were employed. The model consisted of 3141 nodes and 6800 elements.

From the FEA, the stress and strain of the tensile specimens were obtained by averaging these values of the nodes in the reduced section surface ($0 \leq X \leq L_1^c/2$, $0 \leq Y \leq W_1^c/2$, $Z = T^c/2$) because the calculated solutions were different for each node due to the model of randomly oriented CNF-reinforced polymer composites. Additionally, to simulate the randomness of the solutions, an FEA was performed for each three times under the same conditions. Finally, the Young's modulus E^c obtained from the FEA was compared with the experimental results.

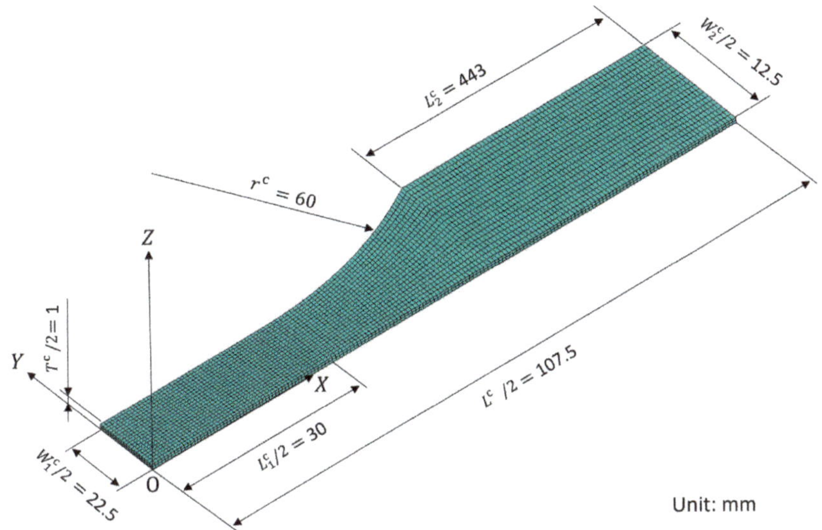

Figure 3. One-eighth finite element model of tensile specimen.

3. Experimental Procedure

In this work, nanocomposites consisting of an epoxy matrix filled with dry CNFs were considered. The diameters of the dry CNFs ranged from 25 to 50 nm, whereas their lengths were between 5 and 50 μm. The processing concept for CNF/epoxy composite is summarized in Figure 4. The fabricated composite samples contained CNF concentrations of 0.73, 2.2, and 3.7 vol %. Neat epoxy samples were also prepared.

Tensile tests were conducted on the CNF/epoxy composite specimen following JIS K 7164, and the stress–strain curves were measured. The stress was determined by dividing the load by the cross-sectional area of the narrow section. Test specimen dimensions are shown in Figure 3. At least five tests were performed under each condition.

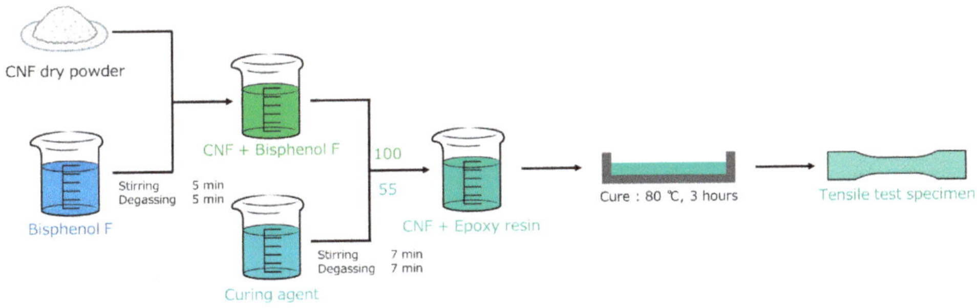

Figure 4. Schematic illustration of the preparation route for CNF/epoxy composite specimen.

4. Results and Discussion

4.1. Parameters of Representative Volume Element

The Young's modulus of the CNF is $E^f = 138$ GPa [7]. The Poisson's ratio v^f was assumed to be equal to that of carbon nanotube and hence $v^f = 0.2$ [17]. On the other hand, the Young's modulus $E^m = 3.23$ GPa and Poisson's ratio $v^m = 0.34$ of the neat epoxy were obtained from the tensile tests.

Figure 5 gives the scanning electron microscopy (SEM) images of the surface for the 3.7 vol % and of fracture surface for the 0.73 vol % CNF/epoxy composite specimen. It was found that the agglomerated CNF clusters were almost uniformly dispersed in the epoxy matrix. The fracture of CNF clusters was observed on the fracture surface of CNF/epoxy composites without disentwining. This result indicates that the applied load was transferred to CNF clusters and that CNF clusters contributed to enhance the epoxy resin matrix during the tensile test. The CNF in the epoxy matrix tended to be agglomerated, and the apparent aspect ratio became small. From this image, the aspect ratio $L^f/d^f = L^m/W^m = 7.5$ was assumed, taking L^f and d^f as the length and diameter of the agglomerated CNF cluster. For an aspect ratio above 50, the predicted Young's modulus is almost independent of the aspect ratio [17]. For simplicity here, the agglomerated CNF clusters were assumed to be isotropic and to have the same Young's modulus and Poisson's ratio as the CNF.

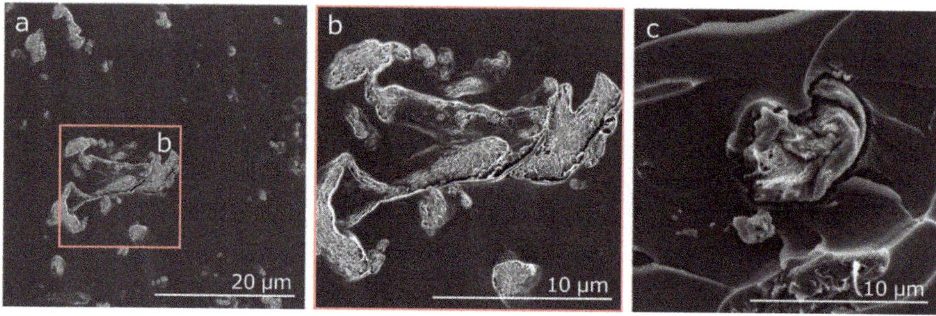

Figure 5. SEM images of (**a**) the surface for the 3.7 vol % CNF/epoxy composite, (**b**) the surface for the 3.7 vol % CNF/epoxy composite (high resolution), and (**c**) the fracture surface for the 0.73 vol % CNF/epoxy composite.

4.2. CNF/epoxy Composite Analysis

Figure 6 shows the measured and calculated stresses as a function of the measured strain for the neat epoxy and 2.2 vol % CNF/epoxy composite. Dotted lines indicate the measured data. Dashed lines are the calculated normal stress in the X-direction at a chosen point ($X = Y = 0$ mm and $Z = T^c/2 = 1$ mm here) from a linear elastic FEA. Each measured curve exhibits the linear initial region, and the calculated curves agree with the measured data very well. In the later loading stage, nonlinearity of the curve exists. This is due to the material nonlinearity. In order to determine the nonlinear stress–strain response numerically, the uniaxial stress–strain behavior, for example, based on Ref. [18] must be represented. More details can be found in our previous works on elastic-plastic finite element modeling [14,19], which is out of scope of the current work. Figure 6 also shows the calculated result for the 2.2 vol % CNF/epoxy composite from the multi-scale FEA (see solid line). It is interesting to note that the stress obtained from the multi-scale FEA was larger than that obtained from the linear FEA. It is anticipated that unexpected high stress is generated in the composite due to the inhomogeneity of the material. The predicted tensile stress σ^c_{XX} distributions obtained from the (a) linear FEA and (b) multi-scale FEA of the randomly oriented CNF/epoxy composite with the CNF volume fraction V^f of 2.2 vol % and the aspect ratio $L^f/d^f = 7.5$ under normal strain of 0.015 are shown in Figure 7. From Figure 7a, the uniform stress distribution was generated at the reduced section from the linear FEA. On the other hand, from Figure 7b, it was found that the non-uniform stress distribution was generated at the reduced section from the multi-scale FEA. The results indicate that if the CNFs were randomly dispersed and completely bonded to the epoxy matrix, the stress of the CNF/epoxy composite would be higher. In other words, the linear analysis may underestimate the stress.

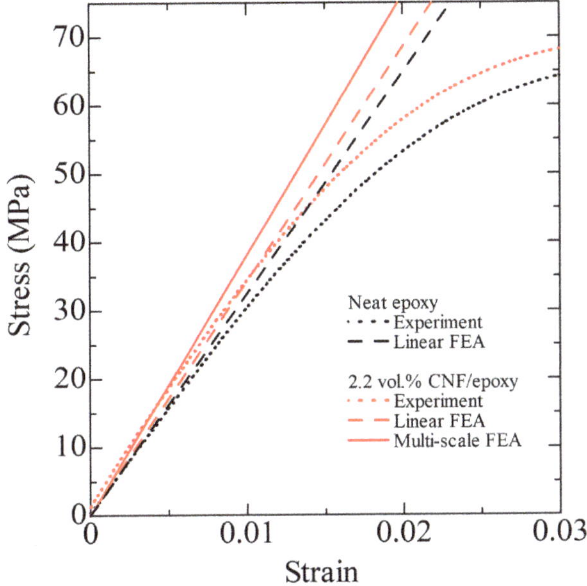

Figure 6. Predicted and experimental stress–strain curves for the neat epoxy and CNF/epoxy composite.

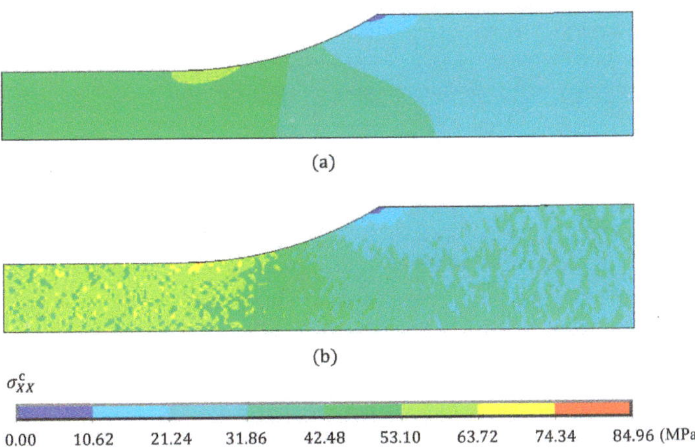

Figure 7. Predicted tensile stress σ^c_{XX} distributions of randomly oriented CNF/epoxy composite under the tensile strain of 0.015 obtained from (**a**) linear FEA and (**b**) multi-scale FEA. The volume fraction is 2.2 vol %, and the aspect ratio is 7.5.

Figure 8 shows the measured and predicted Young's modulus for the neat epoxy, the 2.2 vol % CNF/epoxy composite, and the 3.7 vol % CNF/epoxy composite. The randomness of the solutions was confirmed for each CNF volume fraction in the composites. For the neat epoxy, the predicted Young's modulus agreed with the measured data. On the other hand, the predicted Young's moduli for the CNF/epoxy composites were larger than the measured data. Additionally, although the results are not shown here, when a CNF aspect ratio of 50 was used, the predicted Young's modulus for the 3.7 vol % CNF/epoxy composite became approximately $E^c = 4.65$ GPa. The predicted Young's modulus E^c tends to increase as the CNF volume fraction and aspect ratio increase. This implies that the dispersion of the agglomerated CNF cluster with a high spectral ratio, even in randomly oriented CNF/epoxy composites, increases the Young's modulus.

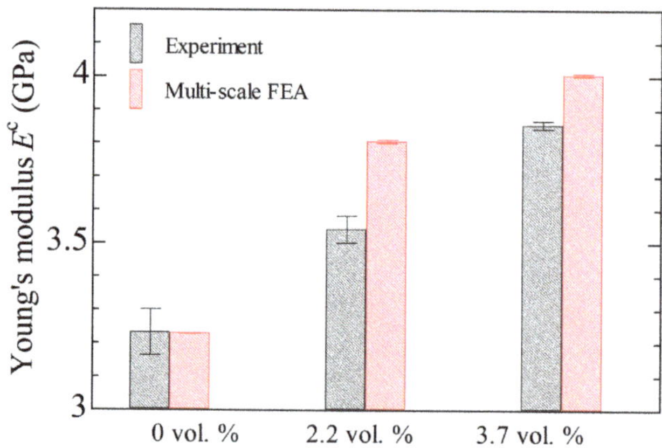

Figure 8. Predicted and experimental Young's modulus for neat epoxy and CNF/epoxy composites.

Table 1 presents a comparison of the measured Young's modulus of the present CNF/epoxy nanocomposite with other nanocomposites from the literature. Although dry CNFs were agglomerated in the epoxy matrix, these CNFs seem to tend to increase the Young's modulus of the epoxy matrix as much as they do other nanomaterials [20–23], except for aligned multi-walled CNT [24]. On the other hand, the flexural modulus was significantly increased by the extremely low CNF slurry addition [11]. This increment cannot be explained by the theories for discontinuous fiber-reinforced composites. Therefore, it seems that the CNF slurry behaves not as a reinforcing fiber in the epoxy matrix but as an auxiliary agent to enhance the structure of the epoxy molecule.

Figure 9 shows the distribution of predicted tensile stress σ_{xx}^{δ} at $y = 0$ plane for the RVE of the composite with the CNF volume fraction V^f of 3.7 vol % and the aspect ratio $L^f/d^f = 7.5$ and 50 under the applied tensile stress of 1 MPa. The stress gradients were confirmed, and the stress increased and became constant toward the center of the CNF. It was found that the stress at the center of the RVE for $L^f/d^f = 50$ was larger than that of $L^f/d^f = 7.5$.

There are many methods to estimate the overall mechanical and physical properties of composites [17,23,24]. In the present work, we assume the perfect bonding between the CNF and epoxy matrix because of its simplicity and use the three-dimensional finite element method. In fact, a weak boundary exists (see Figure 4). Additional modeling is required for a full interpretation of the mechanical and physical properties of CNF/epoxy composites. The main achievements of this research include the development of a finite element model that can be easily extended to complex problems (e.g., CNF-reinforced polymer composites with interfacial layers, CNF-reinforced polymer composites with slip at interfaces).

Table 1. Comparison of the Young's modulus of epoxy matrix nanocomposites.

Nanomaterial	Content	Modulus (GPa)	Increase Rate (%)	
Dry CNF	0	3.23		This study
	0.73 vol %	3.30	2.1	
	2.2 vol %	3.54	10	
	3.7 vol %	3.85	19	
Silica	0	2.96		[20]
	2.5 vol %	3.20	8.1	
	4.9 vol %	3.42	16	
	7.1 vol %	3.57	21	
	9.6 vol %	3.60	22	
	13.4 vol %	3.85	30	
CNT	0	2.90		[21]
	0.1 wt.%	3.01	3.8	
	0.2 wt.%	3.11	7.2	
	0.3 wt.%	3.26	12	
Aligned CNT	0	2.5		[24]
	4.5 vol %	19	660	
	8.4 vol %	32	1180	
	21.4 vol %	50	1900	
Graphene (GP)	0	2.69		[22]
	4.0 wt.%	2.89	7.4	
Surface-modified GP	4.0 wt.%	3.27	22	
Clay	0	3.53		[23]
	2.0 wt.%	3.58	1.4	
	5.0 wt.%	3.66	3.7	
CNF slurry	0	2.20 *		[11]
	0	2.20 *	26	
	0.25 vol %	3.67 *	67	
	0.74 vol %	2.84 *	29	

* Flexural modulus.

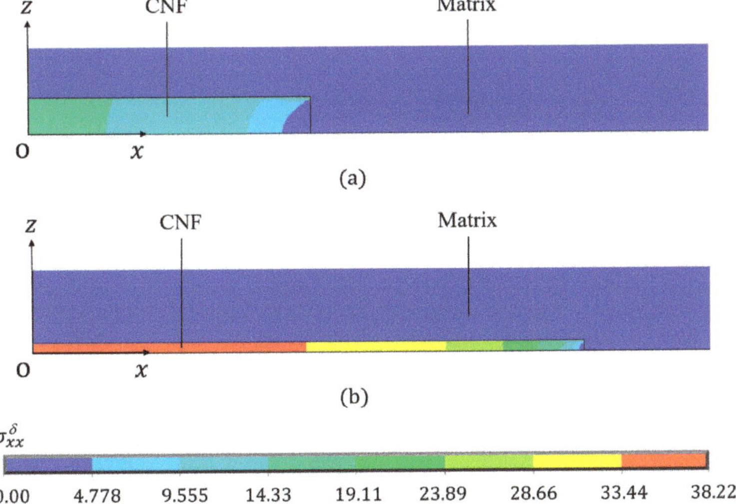

Figure 9. Predicted tensile stress distribution of CNF in RVE with 0.073 vol % under the applied tensile stress of 1 MPa. The aspect ratios are (**a**) 7.5 and (**b**) 50, respectively.

5. Conclusions

In this paper, we studied the tensile behavior of CNF/epoxy composites by performing multi-scale FEA and tensile tests. It was shown that by increasing CNF content, the Young's modulus increases. It was also found that agglomeration of CNF decreases the aspect ratio of CNF and decreases the Young's modulus. A correlation, obtained between the numerical and experimental results for the Young's moduli of the CNF/epoxy composites under tension, revealed that unexpected high stress is generated in CNF/epoxy composites due to the inhomogeneity, and the inhomogeneity tends to increase the Young's modulus. The results from this study offer some basic insights into the reinforcing mechanisms in CNF/epoxy composites under tension. These will enable the design of multifunctional nanocomposite materials based on CNF for advanced engineering applications.

All conclusions are based on the numerical predictions for CNF/epoxy composites with perfect bonding between the CNF and epoxy matrix. Some nanocomposites indicate the presence of weak boundaries. This is a challenging research area, and sooner or later, some progress will be made.

Author Contributions: Conceptualization, F.N.; methodology, Y.W.; software, Y.W.; validation, F.N. and H.K.; formal analysis, M.S.; investigation, F.N. and M.S.; writing—original draft preparation, M.S.; writing—review and editing, F.N.; visualization, H.K.; supervision, F.N.; project administration, F.N. All authors have read and agreed to the published version of the manuscript.

Funding: This research received no external funding.

Conflicts of Interest: The authors declare no conflict of interest.

References

1. Cao, W.-T.; Chen, F.-F.; Zhu, Y.-J.; Zhang, Y.-G.; Jiang, Y.-Y.; Ma, M.-G.; Chen, F. Binary strengthening and toughening of MXene/cellulose nanofiber composite paper with nacre-inspired structure and superior electromagnetic interference shielding properties. *ACS Nano* **2018**, *12*, 4583–4593. [CrossRef] [PubMed]
2. Fazeli, M.; Keley, M.; Biazar, E. Preparation and characterization of starch-based composite films reinforced by cellulose nanofibers. *Int. J. Biol. Macromol.* **2018**, *116*, 272–280. [CrossRef] [PubMed]
3. Nissilä, T.; Hietala, M.; Oksman, K. A method for preparing epoxy–cellulose nanofiber composites with an oriented structure. *Compos. Part A* **2019**, *125*, 105515. [CrossRef]
4. Noguchi, T.; Endo, M.; Niihara, K.; Jinnai, H.; Isogai, A. Cellulose nanofiber/elastomer composites with high tensile strength, modulus, toughness, and thermal stability prepared by high-shear kneading. *Compos. Sci. Technol.* **2020**, *188*, 108005. [CrossRef]
5. Nishino, T.; Takano, K.; Nakamae, K. Elastic modulus of the crystalline regions cellulose polymorphs. *J. Polym. Sci. Part B* **1995**, *33*, 1647. [CrossRef]
6. Bledzki, A.; Gassan, J. Composites reinforced with cellulose based fibres. *Prog. Polym. Sci.* **1999**, *24*, 221–274. [CrossRef]
7. Sain, M.; Oksman, K. Cellulose nanocomposites. *J. Am. Chem. Soc.* **2006**, *938*, 2–8.
8. Chirayil, C.J.; Mathew, L.; Hassan, P.A.; Mozetic, M.; Thomas, S. Rheological behavior of nanocellulose reinforced unsaturated polyester nanocomposites. *Int. J. Biol. Macromol.* **2014**, *69*, 274–281. [CrossRef]
9. Ansari, F.; Skrifvars, M.; Berglund, L. Nanostructured biocomposites based on unsaturated polyester resin and a cellulose nanofiber network. *Compos. Sci. Technol.* **2015**, *117*, 298–306. [CrossRef]
10. Kurita, H.; Yingmei, X.; Katabira, K.; Narita, F. The insert effect of cellulose nanofiber layer on glass fiber-reinforced plastic laminates and their flexural properties. *Mater. Des. Process. Commun.* **2019**, *1*, e58. [CrossRef]
11. Xie, Y.; Kurita, H.; Ishigami, R.; Narita, F. Assessing the flexural properties of epoxy composites with extremely low addition of cellulose nanofiber content. *Appl. Sci.* **2020**, *10*, 1159. [CrossRef]
12. Takeda, T.; Shindo, Y.; Narita, F. Three-dimensional thermoelastic analysis of cracked plain weave glass/epoxy composites at cryogenic temperatures. *Compos. Sci. Technol.* **2004**, *64*, 2353–2362. [CrossRef]
13. Miura, M.; Shindo, Y.; Takeda, T.; Narita, F. Effect of damage on the interlaminar shear properties of hybrid composite laminates at cryogenic temperatures. *Compos. Struct.* **2010**, *93*, 124–131. [CrossRef]

14. Shindo, Y.; Kuronuma, Y.; Takeda, T.; Narita, F.; Fu, S.-Y. Electrical resistance change and crack behavior in carbon nanotube/polymer composites under tensile loading. *Compos. Part B* **2012**, *43*, 39–43. [CrossRef]

15. Levy, A.; Papazian, J.M. Tensile properties of short fiber-reinforced SiC/Al composites: Part II. Finite-element analysis. *Metall. Trans. A* **1990**, *21*, 411–420. [CrossRef]

16. Sun, C.T.; Vaidya, R.S. Prediction of composite properties from a representative volume element. *Compos. Sci. Technol.* **1996**, *56*, 171–179. [CrossRef]

17. Takeda, T.; Shindo, Y.; Narita, F.; Mito, Y. Tensile characterization of carbon nanotube-reinforced polymer composites at cryogenic temperatures: Experiments and multiscale simulations. *Mater. Trans.* **2009**, *50*, 436–445. [CrossRef]

18. Hutchinson, J.W.; Neale, K.W. Neck propagation. *J. Mech. Phys. Solids* **1983**, *31*, 405–426. [CrossRef]

19. Narita, F.; Shindo, Y.; Takeda, T.; Kuronuma, Y.; Sanada, K. Loading rate-dependent fracture properties and electrical resistance-based crack growth monitoring of polycarbonate reinforced with carbon nanotubes under tension. *ASTM J. Test. Eval.* **2015**, *43*, 115–122. [CrossRef]

20. Johnsen, B.B.; Kinloch, A.J.; Mohammed, R.D.; Taylor, A.C.; Sprenger, S. Toughening mechanisms of nanoparticle-modified epoxy polymers. *Polymer* **2007**, *48*, 530–541. [CrossRef]

21. Hsieh, T.H.; Kinloch, A.J.; Taylor, A.C.; Kinloch, I. The effect of carbon nanotubes on the fracture toughness and fatigue performance of a thermosetting epoxy polymer. *J. Mater. Sci.* **2011**, *46*, 7525–7535. [CrossRef]

22. Zaman, I.; Phan, T.T.; Kuan, H.-C.; Meng, Q.L.L.; Bao, T.; Luong, L.; Youssf, O.; Ma, J. Epoxy/graphene platelets nanocomposites with two levels of interface strength. *Polymer* **2011**, *52*, 1603–1611. [CrossRef]

23. Guevara-Morales, A.; Taylor, A.C. Mechanical and dielectric properties of epoxy–clay nanocomposites. *J. Mater. Sci.* **2014**, *49*, 1574–1584. [CrossRef]

24. Ogasawara, T.; Moon, S.-Y.; Inoue, Y.; Shimamura, Y. Mechanical properties of aligned multi-walled carbon nanotube/epoxy composites processed using a hot-melt prepreg method. *Compos. Sci. Technol.* **2011**, *71*, 1826–1833. [CrossRef]

 *nanomaterials*

Article

Buckling Behavior of FG-CNT Reinforced Composite Conical Shells Subjected to a Combined Loading

Abdullah H. Sofiyev [1,*], **Francesco Tornabene** [2], **Rossana Dimitri** [2] **and Nuri Kuruoglu** [3]

[1] Department of Civil Engineering of Engineering Faculty, Suleyman Demirel University, 32260 Isparta, Turkey

[2] Department of Innovation Engineering, University of Salento, 73100 Lecce, Italy; francesco.tornabene@unibo.it (F.T.); rossana.dimitri@unisalento.it (R.D.)

[3] Department of Civil Engineering of Faculty of Engineering and Architecture, Istanbul Gelisim University, 34310 Istanbul, Turkey; nkuruoglu@gelisim.edu.tr

* Correspondence: abdullahavey@sdu.edu.tr; Tel.: +90-246-2111195; Fax: +90-246-2370859

Received: 2 February 2020; Accepted: 26 February 2020; Published: 28 February 2020

Abstract: The buckling behavior of functionally graded carbon nanotube reinforced composite conical shells (FG-CNTRC-CSs) is here investigated by means of the first order shear deformation theory (FSDT), under a combined axial/lateral or axial/hydrostatic loading condition. Two types of CNTRC-CSs are considered herein, namely, a uniform distribution or a functionally graded (FG) distribution of reinforcement, with a linear variation of the mechanical properties throughout the thickness. The basic equations of the problem are here derived and solved in a closed form, using the Galerkin procedure, to determine the critical combined loading for the selected structure. First, we check for the reliability of the proposed formulation and the accuracy of results with respect to the available literature. It follows a systematic investigation aimed at checking the sensitivity of the structural response to the geometry, the proportional loading parameter, the type of distribution, and volume fraction of CNTs.

Keywords: nanocomposites; buckling; FG-CNTRC; truncated cone; critical combined loads

---

1. Introduction

Conical shells are well known to play a key role in many applications, including aviation, rocket and space technology, shipbuilding and automotive, energy and chemical engineering, as well as industrial constructions. In such contexts, carbon nanotubes (CNTs) have increasingly attracted the attention of engineers and designers for optimization purposes, due to their important physical, chemical, and mechanical properties. The shell structures reinforced with CNTs, indeed, are lightweight and resistant to corrosion and feature a high specific strength, with an overall simplification in their manufacturing, transportation, and installation processes.

In many engineering and building structures, shells are subjected to a simultaneous action of different loads, such as a combined compressive force and external pressure, which can affect significantly their global stability, as observed in the pioneering works [1–5], within a parametric study of the buckling response for homogeneous composite cylindrical and conical shells subjected to a combined loading (CL). Among the novel class of composite functionally graded materials (FGMs), the first studies on the buckling response of FGM-based shells subjected to a CL can be found in [6] and [7], for cylindrical and conical shells, as well as in [8–14] for different shell geometries and boundary conditions, while considering different theoretical approaches. Moreover, the increased development of nanotechnology has induced a large adoption of nano-scale materials, e.g., CNTs, in many engineering systems and devices, discovered experimentally by Iijima [15] in 1991 during the production of fullerene by arc discharge evaporation. It is known from the literature, indeed, that the generation of CNTs is strictly related to the creation and evaporation of fullerene, which is decomposed into graphene to yield different types of CNTs. The tubes obtained by graphite with the arc-evaporation process become hollow pipes when the graphite layer, i.e., graphene, turns into a cylindrical shape [16,17]. Improving the properties of materials through a reinforcement phase is one of the most relevant topics in modern materials science (metamaterials, heterogeneous materials, architectured materials etc.) [18–20]. The outstanding mechanical, electrical, and thermal properties of FG CNTs make them very attractive for many current and future engineering applications, more than conventional carbon fiber reinforced composites [21–23]. The modern technology has also allowed a combined use of FGMs and CNTs in various structural elements, which is reflected in the introduction of a great number of advanced theoretical and numerical methods to solve even more complicated problems, with a special focus on mesh-free methods [24–32].

Among the available literature, the formulation and solution of the buckling and postbuckling problems of carbon nanotube reinforced composite (CNTRC)-cylindrical shells under a CL, was introduced for the first time by Shen and Xiang [33], followed by Sahmani et al. [34] for composite nanoshells, including the effect of surface stresses at large displacements, and by the instability study in [35] for rotating FG-CNTRC-cylindrical shells. In the literature, however, many works focusing on the buckling behavior of FG-CNTRC-shells consider the separate action of axial or lateral loads, see [36–44], whereas limited attention has been paid, up to date, to a CL condition. This aspect is considered in the present work for FG-CNTRC-conical shells, whose problem is solved in a closed form through the Galerkin method. A systematic study is performed to evaluate the sensitivity of the buckling response to the geometry, loading condition, distribution, and volume fraction of the reinforcing CNTs, which could be of great interest for design purposes.

The paper is structured as follows: in Section 2 we present the basic formulation of the problem, whose governing equations are presented in Section 3, and solved in closed form in Section 4. The numerical results from the parametric investigation are analyzed in Section 5, while the concluding remarks are discussed in Section 6.

2. Formulation of the Problem

Let us consider an FG-CNTRC truncated conical shell, with length L, half-vertex angle γ, end radii R_1 and R_2 (with $R_1 < R_2$), and thickness h, as schematically depicted in Figure 1, along with the displacement components u, v, and w of an arbitrary point at the reference surface. The FG-CNTRC-conical shells (CSs) are subjected to a combined axial compression and uniform external pressures, as follows:

$$N_{x0} = -T_{ax} - 0.5xP_1 \tan\gamma, \quad N_{\theta 0} = -xP_2 \tan\gamma, \quad N_{x\theta 0} = 0 \tag{1}$$

where $N_{x0}, N_{\theta 0}, N_{x\theta 0}$ are the membrane forces for null initial moments, T_{ax} is the axial compression, and $P_j(j = 1,2)$ stands for the uniform external pressures.

If the external pressures in Figure 1 consider only the lateral pressure, it is $T_{ax} = P_1 = 0$ and $P_2 = P_L$, whereas for a hydrostatic pressure, it is assumed $T_{ax} = 0$ and $P_1 = P_2 = P_H$.

In the following formulation, we consider the volume fraction of CNTs and matrix, denoted by V_{CN} and V_m, respectively, with normal and shear elastic properties $E_{11}^{CN}, E_{22}^{CN}, G_{12}^{CN}$, for CNTs, and E^m, G^m, for the matrix, and efficiency parameters $\eta_j(j = 1,2,3)$ for CNTs. Thus, the mechanical properties of CNTRC-CSs can be expressed, according to an improved mixture rule [33], as follows:

$$E_{11} = \eta_1 V_{CN} E_{11}^{CN} + V_m E^m, \quad \frac{\eta_2}{E_{22}} = \frac{V_{CN}}{E_{22}^{CN}} + \frac{V_m}{E^m}, \quad \frac{\eta_3}{G_{12}} = \frac{V_{CN}}{G_{12}^{CN}} + \frac{V_m}{G^m}, \quad G_{13} = G_{12}, \quad G_{23} = 1.2G_{12}$$

$$\tag{2}$$

where the volume fraction of CNTs and matrix are related as $V_{CN} + V_m = 1$.

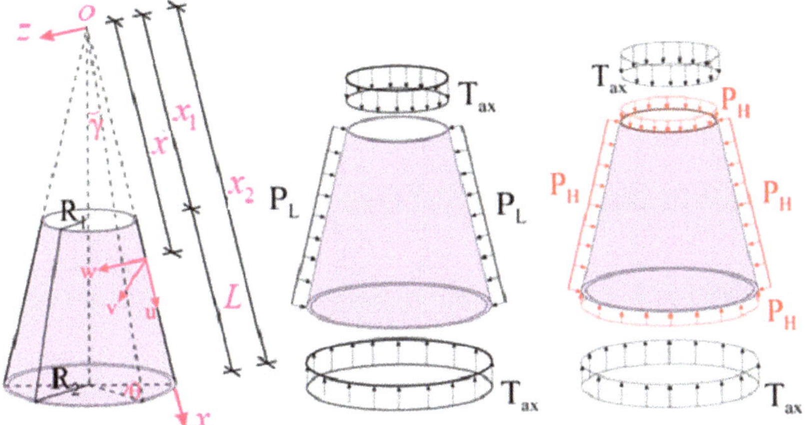

Figure 1. The functionally graded carbon nanotube reinforced composite conical shell (FG-CNTRC-CS) subjected to a combined loading (CL).

The volume fraction of the FG-CNTRC-CS is assumed as follows:

$$V_{CN} = (1 - 2\bar{z})V^*_{CN} \quad \text{for} \ \ FG - V$$

$$V_{CN} = (1 + 2\bar{z})V^*_{CN} \quad \text{for} \ \ FG - \Lambda \tag{3}$$

$$V_{CN} = 4|\bar{z}|V^*_{CN} \quad \text{for} \ \ FG - X, \qquad \bar{z} = z/h$$

where V^*_{CN} is the volume fraction of the CNT, expressed as

$$V^*_{CN} = \frac{w_{CN}}{w_{CN} + (\rho^{CN}/\rho^m) - (\rho^{CN}/\rho^m)w_{CN}} \tag{4}$$

whereby the mass fraction of CNTs is denoted by w_{CN}, and the density of CNTs and matrix are defined as ρ^{CN} and ρ^m, respectively. In our case, for a uniform distribution (UD)-CNTRC-CSs, it is $V_{CN} = V^*_{CN}$. The Poisson's ratio is defined as

$$\mu_{12} = V^*_{CN}\mu^{CN}_{12} + V_m\mu^m \tag{5}$$

The topologies of UD- and FG-CNTRC-CSs are shown in Figure 2.

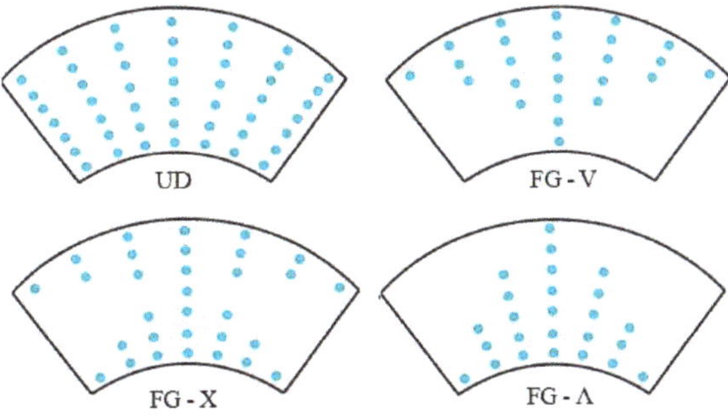

Figure 2. Configurations of uniform distribution (UD)-CNTRC-CSs and three types of FG-CNTRC-CSs.

3. The Governing Equations

Based on the first order shear deformation theory (FSDT), the constitutive stress–strain relations for FG-CNTRC-CSs are expressed as follows:

$$
\begin{bmatrix} \tau_x \\ \tau_\theta \\ \tau_{xz} \\ \tau_{\theta z} \\ \tau_{x\theta} \end{bmatrix}
=
\begin{bmatrix}
\overline{E}_{11}(\overline{z}) & \overline{E}_{12}(\overline{z}) & 0 & 0 & 0 \\
\overline{E}_{21}(\overline{z}) & \overline{E}_{22}(\overline{z}) & 0 & 0 & 0 \\
0 & 0 & \overline{E}_{44}(\overline{z}) & 0 & 0 \\
0 & 0 & 0 & \overline{E}_{55}(\overline{z}) & 0 \\
0 & 0 & 0 & 0 & \overline{E}_{66}(\overline{z})
\end{bmatrix}
\begin{bmatrix} \varepsilon_x \\ \varepsilon_\theta \\ \gamma_{xz} \\ \gamma_{\theta z} \\ \gamma_{x\theta} \end{bmatrix}
\tag{6}
$$

where $\tau_{ij}(i,j = x,\theta,z)$, $\varepsilon_{jj}(j = x,\theta)$, and $\gamma_{ij}(i,j = x,\theta,z)$ are the stress and strain tensors of FG-CNTRC-CSs, respectively, and the coefficients $E_{ij}(\overline{z})$, $(i,j = 1,2,6)$ are defined as

$$
\overline{E}_{11}(\overline{z}) = \frac{E_{11}(\overline{z})}{1-\mu_{12}\mu_{21}}, \overline{E}_{22}(\overline{z}) = \frac{E_{22}(\overline{z})}{1-\mu_{12}\mu_{21}(\overline{z})}
$$

$$
\overline{E}_{12}(\overline{z}) = \frac{\mu_{21}E_{11}(\overline{z})}{1-\mu_{12}\mu_{21}} = \frac{\mu_{12}E_{22}(\overline{z})}{1-\mu_{12}\mu_{21}} = \overline{E}_{21}(\overline{z}),
\tag{7}
$$

$$
\overline{E}_{44}(\overline{z}) = G_{23}(\overline{z}), \overline{E}_{55}(\overline{z}) = G_{13}(\overline{z}), \overline{E}_{66}(\overline{z}) = G_{12}(\overline{z})
$$

The shear stresses of FG-CNTRC-CSs vary throughout the thickness direction as follows [45,46]:

$$
\tau_z = 0, \quad \tau_{xz} = \frac{du_1(z)}{dz}\varphi_1(x,\theta), \quad \tau_{\theta z} = \frac{du_2(z)}{dz}\varphi_2(x,\theta)
\tag{8}
$$

where $\varphi_1(x,\theta)$ and $\varphi_2(x,\theta)$ are the rotations of the reference surface about the θ and x axes, respectively, and $u_1(z)$ and $u_2(z)$ refer to the shear stress shape functions.

By combining Equations (6) and (8), we get the following strain relationships:

$$
\begin{bmatrix} \varepsilon_x \\ \varepsilon_\theta \\ \gamma_{x\theta} \end{bmatrix}
=
\begin{bmatrix}
e_x - z\frac{\partial^2 w}{\partial x^2} + F_1(z)\frac{\partial \varphi_1}{\partial x} \\
e_\theta - z\left(\frac{1}{x^2}\frac{\partial^2 w}{\partial \alpha^2} + \frac{1}{x}\frac{\partial w}{\partial x}\right) + F_2(z)\frac{1}{x}\frac{\partial \varphi_2}{\partial \alpha} \\
\gamma_{0x\theta} - 2z\left(\frac{1}{x}\frac{\partial^2 w}{\partial x \partial \alpha} - \frac{1}{x^2}\frac{\partial w}{\partial \alpha}\right) + F_1(z)\frac{1}{x}\frac{\partial \varphi_1}{\partial \alpha} + F_2(z)\frac{\partial \varphi_2}{\partial x}
\end{bmatrix}
\tag{9}
$$

where $\alpha = \theta \sin\gamma$ and $e_x, e_\theta, \gamma_{0x\theta}$ stand for the strain components at the reference surface, and $F_1(z)$, $F_2(z)$ are defined as

$$
F_1(z) = \int_0^z \frac{1}{\overline{E}_{55}(\overline{z})}\frac{du_1}{dz}dz, \quad F_2(z) = \int_0^z \frac{1}{\overline{E}_{44}(\overline{z})}\frac{du_1}{dz}dz
\tag{10}
$$

The internal actions can be defined in approximate form as follows [46–48]:

$$
\left(N_{ij}, Q_i, M_{ij}\right) = \int_{-h/2}^{h/2} \left(\tau_{ij}, \tau_{iz}, z\tau_{ij}\right)dz, \quad (i,j = x,\theta)
\tag{11}
$$

By introducing the Airy stress function (Φ) satisfying [45,47], Equation (11) becomes as follows:

$$
(N_x, N_\theta, N_{x\theta}) = h\left[\frac{1}{x}\left(\frac{1}{x}\frac{\partial^2}{\partial \alpha^2} + \frac{\partial}{\partial x}\right), \frac{\partial^2}{\partial x^2}, -\frac{1}{x}\left(\frac{\partial^2}{\partial x \partial \alpha} - \frac{1}{x}\frac{\partial}{\partial \alpha}\right)\right]\Phi
\tag{12}
$$

By using Equations (6), (9), (11), and (12), we obtain the expressions for force, moment, and strain components in the reference surface, which are then substituted in the stability and compatibility equations [45,47] to obtain the following governing differential equations for FG-CNTRC-CSs under a CL, with independent parameters Φ, w, φ_1, φ_2, i.e.,

$$L_{11}\Phi + L_{12}w + L_{13}\varphi_1 + L_{14}\varphi_2 = 0$$

$$L_{21}\Phi + L_{22}w + L_{23}\varphi_1 + L_{24}\varphi_2 = 0$$

$$L_{31}\Phi + L_{32}w + L_{33}\varphi_1 + L_{34}\varphi_2 = 0 \tag{13}$$

$$L_{41}\Phi + L_{42}w + L_{43}\varphi_1 + L_{44}\varphi_2 = 0$$

where $L_{ij}(i,j = 1,2,3,4)$ are differential operators, whose details are described in Appendix A.

4. Solution Procedure

The approximating functions for conical shells with free supports are assumed as

$$\Phi = \overline{\Phi}x_2 e^{(a+1)\overline{x}}\sin(n_1\overline{x})\cos(n_2\alpha), \quad w = \overline{w}e^{a\overline{x}}\sin(n_1\overline{x})\cos(n_2\alpha),$$

$$\varphi_1 = \overline{\varphi}_1 e^{a\overline{x}}\cos(n_1\overline{x})\cos(n_2\alpha), \quad \varphi_2 = \overline{\varphi}_2 e^{a\overline{x}}\sin(n_1\overline{x})\sin(n_2\alpha) \tag{14}$$

where $\overline{\Phi}, \overline{w}, \overline{\varphi}_1, \overline{\varphi}_2$ are the unknown constants, a is an unknown coefficient to be determined with the enforcement of the minimum conditions for combined buckling loads, $\overline{x} = \ln\left(\frac{x}{x_2}\right)$, $n_1 = \frac{m\pi}{x_0}$, $n_2 = \frac{n}{\sin\gamma}$, $x_0 = \ln\left(\frac{x_2}{x_1}\right)$, with m and n the wave numbers.

By introducing Equation (14) into Equation (13), and by some manipulation and integration, we determine the nontrivial solution by enforcing

$$\det(c_{ij}) = 0 \tag{15}$$

where $c_{ij}(i,j = 1,2,\ldots,4)$ are the matrix coefficients, as defined in Appendix B.

Equation (15) can be rewritten in expanded form as follows:

$$c_{41}\Gamma_1 - \left(T_{ax}c_T + P_1 c_{P_1} + P_2 c_{P_2}\right)\Gamma_2 + c_{43}\Gamma_3 + c_{44}\Gamma_4 = 0 \tag{16}$$

where c_T is the axial load parameter, c_{P_1} and c_{P_2} are the external pressure parameters, whose expression are given in Appendix B, while parameters $\Gamma_j(j = 1,2,3,4)$ are defined as

$$\Gamma_1 = -\begin{vmatrix} c_{12} & c_{13} & c_{14} \\ c_{22} & c_{23} & c_{24} \\ c_{32} & c_{33} & c_{34} \end{vmatrix}, \quad \Gamma_2 = \begin{vmatrix} c_{11} & c_{13} & c_{14} \\ c_{21} & c_{23} & c_{24} \\ c_{31} & c_{33} & c_{34} \end{vmatrix}, \quad \Gamma_3 = -\begin{vmatrix} c_{11} & c_{12} & c_{14} \\ c_{21} & c_{22} & c_{24} \\ c_{31} & c_{32} & c_{34} \end{vmatrix}, \quad \Gamma_4 = \begin{vmatrix} c_{11} & c_{12} & c_{13} \\ c_{21} & c_{22} & c_{23} \\ c_{31} & c_{32} & c_{33} \end{vmatrix} \tag{17}$$

For FG-CNTRC-CSs under an axial load, it is $P_1 = P_2 = 0$, while T_{1SDT}^{axcr} is defined as follows:

$$T_{1SDT}^{axcr} = \frac{c_{41}\Gamma_1 + c_{43}\Gamma_3 + c_{44}\Gamma_4}{\Gamma_2 E_c h c_T} \tag{18}$$

Differently, for FG-CNTRC-CSs under a uniform lateral pressure, it is $T_{ax} = P_1 = 0$; $P_2 = P_L$, whereas the expression for P_{1SDT}^{Lcr} based on the FSDT is as follows:

$$P_{1SDT}^{Lcr} = \frac{c_{41}\Gamma_1 + c_{43}\Gamma_3 + c_{44}\Gamma_4}{\Gamma_2 E_c c_{P_L}} \tag{19}$$

For FG-CNTRC-CSs under a uniform hydrostatic pressure, it is $T_{ax} = 0, P_1 = P_2 = P_H$, and P_{1SDT}^{Hcr} is defined as

$$P_{1SDT}^{Hcr} = \frac{c_{41}\Gamma_1 + c_{43}\Gamma_3 + c_{44}\Gamma_4}{\Gamma_2 E_c c_{PH}} \tag{20}$$

where c_{PH} is a parameter depending on the hydrostatic pressure, as defined in Appendix B.

For a combined axial load/lateral pressure, and a combined axial load/hydrostatic pressure acting on an FG-CNTRC-CS based on the FSDT, the following relation can be used [47]:

$$\frac{T_1}{T_{1SDT}^{axcr}} + \frac{P_{1L}}{P_{1SDT}^{Lcr}} = 1 \tag{21}$$

and

$$\frac{T_1}{T_{1SDT}^{axcr}} + \frac{P_{1H}}{P_{1SDT}^{Hcr}} = 1 \tag{22}$$

where

$$T_1 = T/E_c h, P_{1L} = P_L/E_c, P_{1H} = P_H/E_c \tag{23}$$

Under the assumptions $T_1 = \eta P_{1L}$ and $T_1 = \eta P_{1H}$, in Equations (21) and (22), we get the following expressions:

$$P_{1SDT}^{Lcbcr} = \frac{T_{1SDT}^{axcr} P_{1SDT}^{Lcr}}{\eta P_{1SDT}^{Lcr} + T_{1SDT}^{axcr}} \tag{24}$$

and

$$P_{1SDT}^{Hcbcr} = \frac{T_{1SDT}^{axcr} P_{1SDT}^{Hcr}}{\eta P_{1SDT}^{Hcr} + T_{1SDT}^{axcr}} \tag{25}$$

where $\eta \geq 0$ is the dimensionless load-proportional parameter.

From Equations (24) and (25), we obtained the expressions for the critical loads T_{1CST}^{axcr}, P_{1CST}^{Lcr}, P_{1CST}^{Hcr}, P_{1CST}^{Lcbcr}, P_{1CST}^{Hcbcr}, while neglecting the shear strains.

5. Results and Discussion

5.1. Introduction

In this section, a poly methyl methacrylate (PMMA) reinforced with (10,10) armchair Single Walled CNTs (SWCNTs) was considered for the numerical investigation. The effective material properties of CNTs and PMMA matrix are reported in Table 1 (see [46]), along with the efficiency parameters for three volume fractions of CNTs. The shear stress quadratic shape functions are distributed as $u_1(z) = u_2(z) = z - 4z^3/3h^2$ [46]. The critical CL values for FG-CNTRC-CSs are determined for different magnitudes of a, within a coupled stress theory (CST) context, in order to check for the effect of the FSDT on the critical loading condition. After a systematic numerical computation, it is found that for freely supported FG-CNTRC-CSs, the critical values of a CL were reached for $a = 2.4$.

Table 1. Properties of CNTs and matrix.

	CWCNT	Matrix (PMMA)
Geometrical properties	$\widetilde{L} = 9.26$ nm, $\widetilde{r} = 0.68$ nm, $\widetilde{h} = 0.067$ nm	
Material properties	$E_{11}^{CN} = 5.6466$ TPa, $E_{22}^{CN} = 7.0800$ TPa $G_{12}^{CN} = 1.9445$ TPa, $\mu_{12}^{CN} = 0.175; \rho^{CN} = 1400$ kg/m^3	$E^m = 2.5$ Pa, $\mu^m = 0.34,$ $\rho^m = 1150$ kg/m^3
CNT efficiency parameter	$\eta_1 = 0.137, \eta_2 = 1.022, \eta_3 = 0.715$ for $V_{CN}^* = 0.12;$ $\eta_1 = 0.142, \eta_2 = 1.626, \eta_3 = 1.138$ for $V_{CN}^* = 0.17;$ $\eta_1 = 0.141, \eta_2 = 1.585, \eta_3 = 1.109$ for $V_{CN}^* = 0.28.$	

5.2. Comparative Evaluation

As a first comparative check, the critical lateral pressure, axial load, and combined load of shear deformable CNTRC-CYLSs with an FG-X profile is evaluated as in [33]. The CNTRC-CS reverts to a CNTRC-CYLS, when γ tends to zero. The CNTRC-CYLS has radius r and length L_1 with the following geometrical properties: $r/h = 30$, $h = 2$ mm, and $L_1 = \sqrt{300rh}$, whereas the material properties are assumed as in Table 1, for $T = 300$ K, see [33]. The magnitudes of the critical loading for CNTRC-CYLSs were obtained for $a = 0$. Based on Table 2, a good agreement between our results and predictions by Shen and Xiang [33] is observable for the critical lateral pressure, axial load, and CL.

Table 2. Comparative response of shear deformable CNTRC-CYLSs with the FG-X profile under a separate or combined axial load and lateral pressure.

	T_{SDT}^{axcr} **(MPa)**	P_{SDT}^{Lcr} **(MPa)**	P_{SDT}^{Lcbcr} **(MPa)**	
			$\eta = 750$	$\eta = 140$
V_{CN}^{*}		Shen and Xiang [33]		
0.12	118.848	0.285	0.112	0.218
0.17	196.376	0.484	0.190	0.370
0.28	247.781	0.616	0.242	0.470
		Present study		
0.12	117.840	0.281	0.111	0.2181
0.17	197.515	0.479	0.188	0.3711
0.28	247.062	0.613	0.2414	0.4756

In Table 3, the values of P_{SDT}^{Lcr} of FG-CNTRC-CSs with different profiles and half-vertex angles are compared with results by [37], based on the GDQ method. A FG-CNTRC is considered for $V_{CN}^{*} = 0.17$, $L = \sqrt{300R_1h}$, $R_1/h = 100$, and $h = 1$ mm. Based on Table 3, the correspondence between our values of P_{SDT}^{Lcr} and predictions by Jam and Kiani [37], verifies the consistency of our formulation.

Table 3. Comparative response of shear deformable CNTRC-CSs with the different profiles under lateral pressure for a different half-vertex angle.

		P_{SDT}^{Lcr} (in kPa), (n_{cr})		
	γ	$10°$	$20°$	$30°$
Jam and Kiani [37]	UD	31.11(8)	24.31(9)	19.00(9)
	FG-X	34.53(8)	27.24(9)	21.38(9)
	FG-V	32.41(8)	25.19(9)	19.52(10)
Present study	UD	31.01(8)	23.91(9)	18.23(10)
	FG-X	34.38(8)	26.69(9)	20.49(10)
	FG-V	32.40 (8)	24.97 (9)	19.77 (10)

5.3. Analysis of Combined Buckling Loads

In what follows, we analyze the sensitivity of the critical loading to FG profiles, volume fractions of CNT, and FSDT formulation, by considering the ratios $100\% \times \dfrac{P_{FG-CN}^{cbcr} - P_{UD}^{cbcr}}{P_{UD}^{cbcr}}$ and $100\% \times \dfrac{P_{1CST}^{cbcr} - P_{1SDT}^{cbcr}}{P_{1CST}^{cbcr}}$. One of the main parameters affecting a critical CL is represented by the load-proportional parameter. In Figures 3 and 4, we plot the variations of P_{1SDT}^{Lcbcr}, P_{1CST}^{Lcbcr}, as shown in Figure 3, P_{1SDT}^{Hcbcr} and P_{1CST}^{Hcbcr}, as shown in Figure 4, vs. the dimensionless load-proportional parameter η, for UD- and FG-CNTRCs. Based on both figures, for all profiles, the magnitudes of the critical CL decrease for an increasing dimensionless load-proportional parameter η. This sensitivity is more pronounced for a FG $-\Lambda$ profile, compared to a FG-V or FG-X profile. The strong influence of the FG profiles, V_{CN}, and shear strains on the critical CLs depends on the dimensionless load-proportional parameter.

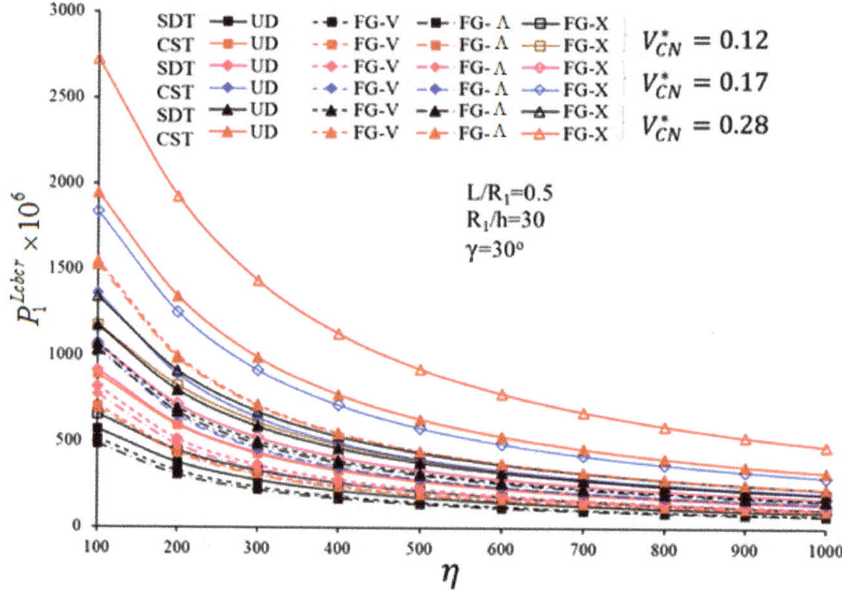

Figure 3. Variation of P_{1SDT}^{Lcbcr} and P_{1CST}^{Lcbcr} for UD- and FG-CNTRC-CSs with different V_{CN}^{*} and load-proportional parameter η.

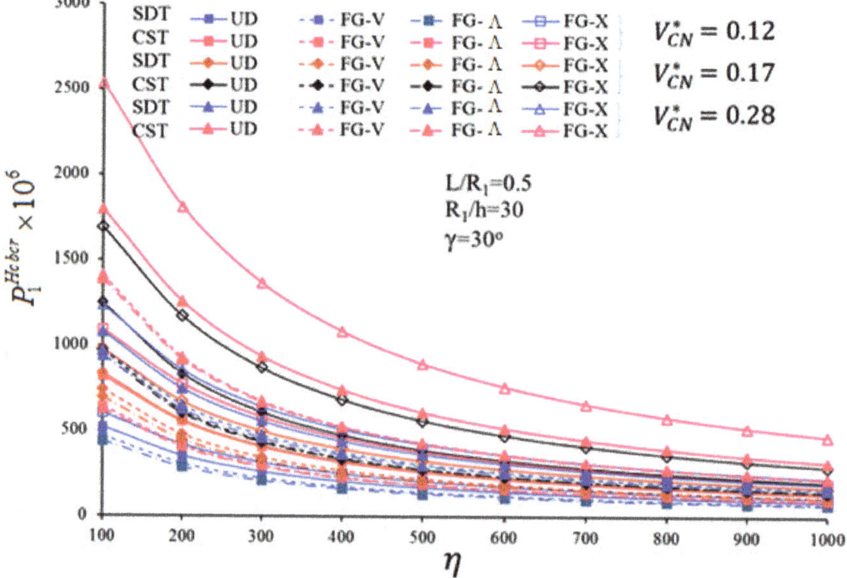

Figure 4. Variation of P_{1SDT}^{Hcbcr} and P_{1CST}^{Hcbcr} for UD- and FG-CNTRC-CSs with a different V_{CN}^{*} and load-proportional parameter η.

A pronounced sensitivity of P_{1SDT}^{Lcbcr} and P_{1SDT}^{Hcbcr} to the FG-CNTRC types, was noticed for η ranging between 100 and 800, whereas a certain influence was observed in a horizontal sense, as $\eta > 800$, for a fixed V_{CN}^*. This last phenomenon can be explained by the fact that, for large values of η, the axial load prevails over the external pressure. For a fixed value of $V_{CN}^* = 0.12$, the effect of a FG profile (i.e., FG-V, FG $- \Lambda$, FG-X, respectively) on the P_{1SDT}^{Lcbcr} is estimated as (-10%), (-14.9%), $(+14.4\%)$ for $\eta = 100$; (-17%), (-21.4%), $(+21.8\%)$ for $\eta = 800$; (-17.04%), (-21.4%), $(+21.9\%)$ for $\eta = 1000$. Compared to a uniform distribution (UD), the highest sensitivity of P_{1SDT}^{Lcbcr} is noticed for $V_{CN}^* = 0.28$ in the FG-X profile, whereby a FG-V type distribution features the lowest sensitivity for the same fixed value of $V_{CN}^* = 0.28$.

A small influence of the shear strain is observable for an increasing value of η and a FG-X profile. This influence continues to decrease by almost 2%–3%, for a FG-V and FG $- \Lambda$ profile with different V_{CN}^*, while reaching the lowest percentage at $V_{CN}^* = 0.17$ for a FG $- \Lambda$ profile.

Table 4 summarizes the variation of the critical CLs, based on a FSDT and CST, for a FG-V and FG-X profile, and different R_1/h ratios. The geometrical data for numerical computations are provided in the same table. Please note that the critical CL decreases monotonically, along with a gradual increase of the circumferential wave numbers, for an increased value of R_1/h.

An irregular effect of the FG-V and FG-X profile on the P_{1SDT}^{Lcbcr} and P_{1SDT}^{Hcbcr} is observed with R_1/h, for a fixed V_{CN}^*. When the dimensionless R_1/h ratio increases from 30 up to 90, the influence of an FG-V profile on the CL values tends to decrease, whereas the effect of a FG-X profile on the CL increases, for an increasing value of R_1/h from 30 to 70. After this value, the effect decreases slightly. The shear strains reduce significantly the influence of FG-V and FG-X profiles on both CLs. For example, the effect of a FG-V profile on P_{1SDT}^{Lcbcr} is less pronounced than P_{1CST}^{Lcbcr} by about 3%–12%. This difference becomes meaningful and varies from 3% up to 23% for a FG-X profile, depending on the selected R_1/h ratio. Note that the highest FG effect on the P_{1SDT}^{Lcbcr} occurs for a FG-X profile $(+35.47\%)$, at $R_1/h = 90$ and $V_{CN}^* = 0.28$, while the lowest effect (-11.69%) is found for a FG-V profile, at $R_1/h = 90$ and $V_{CN}^* = 0.17$. These effects are more pronounced for a combined axial load and hydrostatic pressure (P_{1SDT}^{Hcbcr}), compared to a combined axial load and lateral pressure (P_{1SDT}^{Lcbcr}). The influence of the shear strain on P_{1SDT}^{Lcbcr} for an FG-X and FG-V profile decreases for each fixed value of V_{CN}^*, and remains significant for an increased value of R_1/h up to 90. A similar sensitivity to the shear strain was observed for P_{1SDT}^{Hcbcr}, but was less pronounced than the one for P_{1SDT}^{Lcbcr}. The highest shear strain effect on P_{1SDT}^{Lcbcr} is observed for an FG-X profile (-52.8%), at $R_1/h = 30$ and $V_{CN}^* = 0.28$, whereas the lowest shear strain influence on P_{1SDT}^{Hcbcr} is noticed for an FG-V profile $(+1.54\%)$, at $R_1/h = 70$ and $V_{CN}^* = 0.28$. An increased value of V_{CN}^* yields an irregular effect of the shear strains on the critical CL.

The variation of the critical CL for UD- and FG-CNTRC-CSs with different profiles is plotted in Figures 5 and 6, vs. γ. It seems that the critical value of the CL decreases for an increased value of γ. As γ increases from $15°$ to $60°$, the effect of FG-V and FG-X distributions on the critical magnitude of the CL decreases slightly, whereby it decreases rapidly as $\gamma > 60°$ for all values of V_{CN}^*. Furthermore, the effect of CNT distribution on P_{1SDT}^{Lcbcr} and P_{1SDT}^{Hcbcr} maintains almost the same for different γ. The shear strain effect on the critical CL depends on the selected CNT profile, especially for a FG-X profile. A remarkable shear strain influence of $(+55.56\%)$ on the critical value of the CL occurs at $V_{CN}^* = 0.28$ and $\gamma = 75°$.

Table 4. Variation of the critical CLs for UD- and FG-CNTRC-CSs based on first order shear deformation theory (FSDT) and coupled stress theory (CST) with different V_{CN}^* and R_1/h ratios. $L/R_1 = 0.5$, $\gamma = 30°, \eta = 500$.

V_{CN}^*	R_1/h	Types	P_{1SDT}^{cbLcr} (n_{cr})	P_{1CST}^{cbLcr} (n_{cr})	P_{1SDT}^{cbHcr} (n_{cr})	P_{1CST}^{cbHcr} (n_{cr})
0.12	30	UD	178.227(8)	273.320(7)	172.051(8)	263.619(6)
		FGV-V	149.371(7)	196.380(5)	144.088(7)	189.196(5)
		FGV-X	215.601(9)	393.913(8)	208.297(9)	380.032(7)
	50	UD	83.296(9)	96.267(9)	78.996(9)	91.297(9)
		FGV-V	68.684(8)	72.655(8)	65.013(8)	68.772(8)
		FGV-X	107.856(10)	134.647(10)	102.492(10)	127.951(10)
	70	UD	44.157(11)	46.897(11)	41.493(11)	44.068(11)
		FGV-V	37.515(10)	37.035(10)	35.124(9)	34.636(9)
		FGV-X	57.660(12)	63.714(12)	54.360(12)	60.067(12)
	90	UD	26.050(12)	26.743(12)	24.320(12)	24.967(12)
		FGV-V	22.882(11)	21.902(11)	21.266(11)	20.355(11)
		FGV-X	33.770(13)	35.488(13)	31.663(13)	33.274(13)
0.17	30	UD	277.753(7)	403.421(6)	267.928(7)	388.892(6)
		FGV-D	216.472(7)	285.441(6)	208.815(7)	275.161(6)
		FGV-X	337.997(8)	582.930(7)	326.283(8)	562.312(7)
	50	UD	127.380(9)	144.108(9)	120.804(9)	136.614(8)
		FGV-V	104.824(8)	108.979(8)	99.222(8)	103.074(7)
		FGV-X	166.556(10)	202.672(9)	157.999(9)	192.209(9)
	70	UD	67.718(10)	71.100(10)	63.421(10)	66.589(10)
		FGV-V	57.633(9)	56.330(9)	53.797(9)	52.580(9)
		FGV-X	89.216(11)	97.309(11)	83.834(11)	91.439(11)
	90	UD	40.167(12)	40.992(12)	37.468(11)	38.204(11)
		FGV-V	35.470(11)	33.809(11)	32.965(11)	31.366(10)
		FGV-X	52.637(13)	54.879(13)	49.194(12)	51.292(12)
0.28	30	UD	379.351(8)	632.214(7)	366.204(8)	609.852(7)
		FGV-V	319.603(7)	446.857(6)	308.298(7)	430.763(6)
		FGV-X	437.545(9)	927.004(8)	422.721(9)	894.878(8)
	50	UD	182.000(10)	218.225(10)	172.950(10)	207.373(10)
		FGV-V	148.010(9)	162.024(8)	140.108(8)	153.364(8)
		FGV-X	234.684(10)	314.925(11)	223.013(10)	299.268(10)
	70	UD	96.385(12)	104.268(12)	90.694(11)	98.163(11)
		FGV-V	79.650(10)	80.897(10)	74.597(10)	75.765(10)
		FGV-X	128.856(12)	148.054(12)	121.480(12)	139.579(12)
	90	UD	56.322(13)	58.452(13)	52.808(13)	54.805(13)
		FGV-V	47.930(12)	47.203(11)	44.628(11)	43.870(11)
		FGV-X	76.299(13)	82.031(14)	71.538(13)	76.978(13)

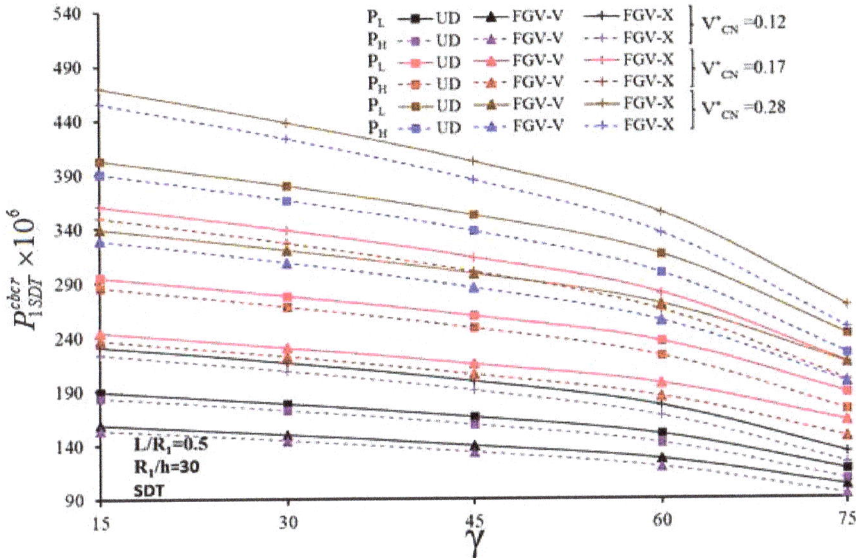

Figure 5. Variation of CCLs for UD- and FG-CNTRC-CSs based on the FSDT with a different V^*_{CN} and half-vertex angle γ, $\eta = 500$.

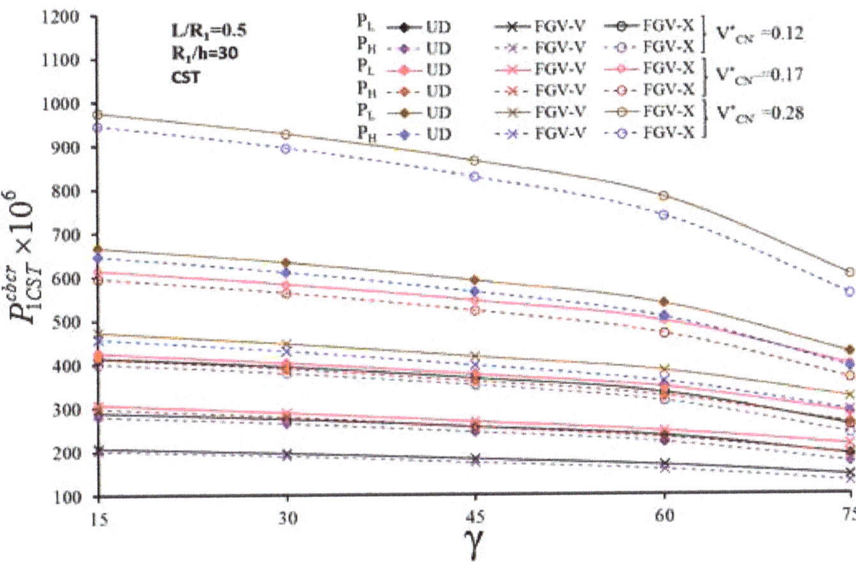

Figure 6. Variation of CCLs for UD- and FG-CNTRC-CSs based on the CST with a different V^*_{CN} and half-vertex angle γ, $\eta = 500$.

6. Conclusions

The buckling of FG-CNTRC-CSs subjected to a combined loading was here studied based on a combined Donnell-type shell theory and FSDT. The FG-CNTRC-CS properties were assumed to vary gradually in the thickness direction with a linear distribution of the volume fraction V_{CN} of CNTs. The governing equations were converted into algebraic equations using the Galerkin procedure, and the analytical expression for the critical value of the combined loading was found. The solutions were compared successfully with results in the open literature, thus confirming the accuracy of the proposed formulation. A novel buckling analysis was, thus, performed for both a uniform distribution and FG distribution of CNTs, while determining the effect of the volume fraction and shell geometry on the critical value of the combined loading condition, as useful for practical engineering applications.

Author Contributions: Conceptualization, A.H.S., N.K., R.D. and F.T.; Formal analysis, A.H.S., N.K., R.D. and F.T.; Investigation, A.H.S., N.K. and F.T.; Validation, N.K., R.D. and F.T.; Writing—Original Draft, A.H.S., N.K., R.D. and F.T.; Writing—Review & Editing, R.D. and F.T. All authors have read and agreed to the published version of the manuscript.

Funding: This research received no external funding.

Conflicts of Interest: The authors declare no conflict of interest.

Appendix A

More details for $L_{ij}(i, j = 1, 2, \ldots, 4)$ differential operators are here defined as follows:

$$
\begin{aligned}
L_{11} = {}& t_{12}h\frac{\partial^4}{\partial x^4} + \frac{(t_{11}-t_{31})h}{x^2}\frac{\partial^4}{\partial x^2\partial\alpha^2} + \frac{(3t_{31}-3t_{11}-t_{21})h}{x^3}\frac{\partial^3}{\partial x\partial\alpha^2} + \frac{(t_{11}-t_{22}+t_{12})h}{x}\frac{\partial^3}{\partial x^3} \\
& + \frac{(t_{22}-t_{11}-t_{12}-t_{21})h}{x^2}\frac{\partial^2}{\partial x^2} + \frac{3(t_{21}+t_{11}-t_{31})h}{x^4}\frac{\partial^2}{\partial\alpha^2} + \frac{2t_{21}h}{x^3}\frac{\partial}{\partial x}, \\
L_{12} = {}& -t_{13}\frac{\partial^4}{\partial x^4} - \frac{t_{14}+t_{32}}{x^2}\frac{\partial^4}{\partial x^2\partial\alpha^2} + \frac{3t_{14}+3t_{32}+t_{24}}{x^3}\frac{\partial^3}{\partial x\partial\alpha^2} - \frac{t_{13}+t_{14}-t_{23}}{x}\frac{\partial^3}{\partial x^3} \\
& + \frac{t_{13}+t_{14}-t_{23}+t_{24}}{x^2}\frac{\partial^2}{\partial x^2} - \frac{3(t_{14}+t_{24}+t_{32})}{x^4}\frac{\partial^2}{\partial\alpha^2} - \frac{2t_{24}}{x^3}\frac{\partial}{\partial x}, \\
L_{13} = {}& t_{15}\frac{\partial^3}{\partial x^3} + \frac{t_{15}-t_{25}}{x}\frac{\partial^2}{\partial x^2} + \frac{t_{35}}{x^2}\frac{\partial^3}{\partial x\partial\alpha^2} - F_3\frac{\partial}{\partial x} - \frac{t_{15}-t_{25}}{x^2}\frac{\partial}{\partial x} - \frac{t_{35}}{x^3}\frac{\partial^2}{\partial\alpha^2}, \\
L_{14} = {}& \frac{t_{38}+t_{18}}{x}\frac{\partial^3}{\partial x^2\partial\alpha} - \frac{t_{28}+t_{18}+t_{38}}{x^2}\frac{\partial^2}{\partial x\partial\alpha} + \frac{2t_{28}}{x^3}\frac{\partial}{\partial\alpha},
\end{aligned}
\tag{A1}
$$

$$
\begin{aligned}
L_{21} = {}& \frac{t_{21}h}{x^3}\frac{\partial^4}{\partial\alpha^4} + \frac{(t_{22}-t_{31})h}{x}\frac{\partial^4}{\partial x^2\partial\alpha^2} + \frac{t_{21}h}{x^2}\frac{\partial^3}{\partial x\partial\alpha^2}, \\
L_{22} = {}& -\frac{t_{32}+t_{23}}{x}\frac{\partial^4}{\partial x^2\partial\alpha^2} - \frac{t_{24}}{x^3}\frac{\partial^4}{\partial\alpha^4} - \frac{t_{24}}{x^2}\frac{\partial^3}{\partial x\partial\alpha^2}, \\
L_{23} = {}& \frac{t_{25}+t_{35}}{x}\frac{\partial^3}{\partial x\partial\alpha^2} + \frac{t_{35}}{x^2}\frac{\partial^2}{\partial\alpha^2}, \\
L_{24} = {}& t_{38}\frac{\partial^3}{\partial x^2\partial\alpha} + \frac{2t_{38}}{x}\frac{\partial^2}{\partial x\partial\alpha} + \frac{t_{28}}{x^2}\frac{\partial^3}{\partial\alpha^3} - F_4\frac{\partial}{\partial\alpha},
\end{aligned}
\tag{A2}
$$

$$
\begin{aligned}
L_{31} = {}& \frac{q_{11}h}{x^4}\frac{\partial^4}{\partial\alpha^4} + \frac{(2q_{31}+q_{21}+q_{12})h}{x^2}\frac{\partial^4}{\partial x^2\partial\alpha^2} - \frac{2(q_{31}+q_{21})h}{x^3}\frac{\partial^3}{\partial x\,\partial\alpha^2} \\
& + \frac{2(q_{31}+q_{21}+q_{11})h}{x^4}\frac{\partial^2}{\partial\alpha^2} + \frac{q_{11}h}{x^3}\frac{\partial}{\partial x} - \frac{q_{11}h}{x^2}\frac{\partial^2}{\partial x^2} + \frac{(q_{21}+2q_{22}-q_{12})h}{x}\frac{\partial^3}{\partial x^3} + q_{22}h\frac{\partial^4}{\partial x^4}, \\
L_{32} = {}& -\frac{q_{14}}{x^4}\frac{\partial^4}{\partial\alpha^4} + \frac{2q_{32}-q_{13}-q_{24}}{x^2}\frac{\partial^4}{\partial x^2\partial\alpha^2} + \frac{2(q_{24}-q_{32})}{x^3}\frac{\partial^3}{\partial x\partial\alpha^2} + \frac{2(q_{32}-q_{24}-q_{14})}{x^4}\frac{\partial^2}{\partial\alpha^2} \\
& - \frac{q_{14}}{x^3}\frac{\partial}{\partial x} + \left(\frac{q_{14}}{x^2} + \frac{\cot\gamma}{x}\right)\frac{\partial^2}{\partial x^2} + \frac{q_{13}-q_{24}-2q_{23}}{x}\frac{\partial^3}{\partial x^3} - q_{23}\frac{\partial^4}{\partial x^4}, \\
L_{33} = {}& \frac{2q_{35}+q_{15}}{x^2}\frac{\partial^3}{\partial x\partial\alpha^2} + q_{25}\frac{\partial^3}{\partial x^3} + \frac{2q_{25}-q_{15}}{x}\frac{\partial^2}{\partial x^2}, \\
L_{34} = {}& \frac{q_{18}}{x^3}\frac{\partial^3}{\partial\alpha^3} + \frac{2q_{38}+q_{28}}{x}\frac{\partial^3}{\partial x^2\partial\alpha} + \frac{2q_{38}-q_{18}}{x^2}\frac{\partial^2}{\partial x\partial\alpha} + \frac{q_{18}}{x^3}\frac{\partial}{\partial\alpha},
\end{aligned}
\tag{A3}
$$

$$
\begin{aligned}
L_{41} &= \frac{h}{x\tan\gamma}\frac{\partial^2}{\partial x^2}, \quad L_{42} = -x\tan\gamma\left[P_1\frac{\partial^2}{\partial x^2} + P_2\left(\frac{1}{x}\frac{\partial^2}{\partial\alpha^2} + \frac{\partial}{\partial x}\right)\right], \\
L_{43} &= F_3\left(\frac{\partial}{\partial x} + \frac{1}{x}\right), \quad L_{44} = \frac{F_4}{x}\frac{\partial}{\partial\alpha}.
\end{aligned}
\tag{A4}
$$

where $F_3 = F_4 = f\left(\frac{h}{2}\right) - f\left(-\frac{h}{2}\right)$ and the following definitions are assumed:

$$
\begin{aligned}
t_{11} &= k_{11}^1 q_{11} + k_{12}^1 q_{21}, \ t_{12} = k_{11}^1 q_{12} + k_{12}^1 q_{21}, \ t_{13} = k_{11}^1 q_{13} + k_{12}^1 q_{23} + k_{11}^2 \\
t_{14} &= k_{11}^1 q_{14} + k_{12}^1 q_{24} + k_{12}^2, \ t_{15} = k_{11}^1 q_{15} + k_{12}^1 q_{25} + k_{15}^1, \ t_{18} = k_{11}^1 q_{18} + k_{12}^1 q_{28} + k_{18}^1, \\
t_{21} &= k_{11}^1 q_{11} + k_{22}^1 q_{21}, \ t_{22} = k_{22}^1 q_{12} + k_{12}^1 q_{22}, \ t_{23} = k_{21}^1 q_{13} + k_{22}^1 q_{23} + k_{21}^1, \\
t_{24} &= k_{22}^1 q_{14} + k_{22}^1 q_{24} + k_{22}^2, \ t_{25} = k_{21}^1 q_{15} + k_{22}^1 q_{25} + k_{25}^1, \ t_{28} = k_{21}^1 q_{18} + k_{22}^1 q_{28} + k_{28}^1, \\
t_{31} &= k_{66}^1 q_{31}, \ t_{32} = k_{66}^1 q_{32} + 2k_{66}^2, \ t_{35} = k_{35}^1 - k_{66}^1 q_{35}, \ t_{38} = k_{38}^1 - k_{66}^1 q_{38}, \\
q_{11} &= \frac{k_{22}^0}{\Pi}; \ q_{12} = -\frac{k_{12}^0}{\Pi}, \ q_{13} = \frac{k_{12}^0 k_{21}^1 - k_{11}^1 k_{22}^0}{\Pi}, \ q_{14} = \frac{k_{12}^0 k_{22}^1 - k_{12}^1 k_{22}^0}{\Pi}, \\
q_{15} &= \frac{k_{25}^0 k_{12}^0 - k_{15}^0 k_{22}^0}{\Pi}, \ q_{18} = \frac{k_{28}^0 k_{12}^0 - k_{18}^0 k_{22}^0}{\Pi}, \ q_{21} = -\frac{k_{21}^0}{\Pi}; \ q_{22} = \frac{k_{11}^0}{\Pi}, \\
q_{23} &= \frac{k_{11}^1 k_{21}^0 - k_{21}^1 k_{11}^0}{\Pi}, \ q_{24} = \frac{k_{12}^1 k_{21}^0 - k_{22}^1 k_{11}^0}{\Pi}, \ q_{25} = \frac{k_{15}^0 k_{21}^0 - k_{25}^0 k_{11}^0}{\Pi}, \ q_{31} = \frac{1}{k_{66}^0}, \\
q_{28} &= \frac{k_{18}^0 k_{21}^0 - k_{28}^0 k_{11}^0}{\Pi}, \ \Pi = k_{11}^0 k_{22}^0 - k_{12}^0 k_{21}^0, \ q_{32} = -\frac{2k_{66}^1}{k_{66}^0}, \ q_{35} = \frac{k_{35}^0}{k_{66}^0}, \ q_{38} = \frac{k_{38}^0}{k_{66}^0}
\end{aligned}
$$

$$\text{(A5)}$$

with

$$
\begin{aligned}
k_{11}^i &= \int_{-h/2}^{h/2} \overline{E}_{11}(\overline{z}) z^i \, \mathrm{d}z, \ k_{12}^i = \int_{-h/2}^{h/2} \overline{E}_{12}(\overline{z}) z^i \, \mathrm{d}z = \int_{-h/2}^{h/2} \overline{E}_{21}(\overline{z}) z^i \, \mathrm{d}z = k_{21}^i, \\
k_{22}^i &= \int_{-h/2}^{h/2} \overline{E}_{22}(\overline{z}) z^i \, \mathrm{d}z, \ k_{66}^i = \int_{-h/2}^{h/2} \overline{E}_{66}(\overline{z}) z^i \, \mathrm{d}z, \ i = 0,1,2. \\
k_{15}^{i_1} &= \int_{-h/2}^{h/2} z^{i_1} F_1(z) \overline{E}_{11}(\overline{z}) \, \mathrm{d}z, \ k_{18}^{i_1} = \int_{-h/2}^{h/2} z^{i_1} F_2(z) \overline{E}_{12}(\overline{z}) \, \mathrm{d}z, \\
k_{25}^{i_1} &= \int_{-h/2}^{h/2} z^{i_1} F_1(z) \overline{E}_{21}(\overline{z}) \, \mathrm{d}z, \ k_{28}^{i_1} = \int_{-h/2}^{h/2} z^{i_1} F_2(z) \overline{E}_{22}(\overline{z}) \, \mathrm{d}z, \\
k_{35}^{i_1} &= \int_{-h/2}^{h/2} z^{i_1} F_1(z) \overline{E}_{66}(\overline{z}) \, \mathrm{d}z, \ k_{38}^{i_1} = \int_{-h/2}^{h/2} z^{i_1} F_2(z) \overline{E}_{66}(\overline{z}) \, \mathrm{d}z, \ i_1 = 0,1.
\end{aligned}
$$

$$\text{(A6)}$$

Appendix B

The parameters $c_{ij} (i, j = 1, 2, 3, 4)$ are defined as follows:

$$
\begin{aligned}
c_{11} = &-\frac{n_1^2 \vartheta_{a-0.5}}{4x_2^3} \left\{ t_{12} \left[3(a-1)(a+1)^3 + 2n_1^2(a+4)(a+1) - n_1^4 \right] - (t_{11} - t_{31}) n_2^2 \left(a^2 - a - 2 + n_1^2 \right) \right\} \\
&-\frac{n_1^2 \vartheta_{a-0.5}}{8x_2^3} \left\{ 3(2t_{31} + t_{21} - 3t_{11}) n_2^2 + (t_{11} - 5t_{12} - t_{22}) \left[(a+1)^2(4a-5) + n_1^2(4a+7) \right] \right. \\
&\left. + 2(7t_{12} + 4t_{22} - 4t_{11} - t_{21}) \left[(a^2 - a - 2) + n_1^2 \right] - 9(t_{11} - t_{12} - t_{22} + t_{21}) \right\}
\end{aligned}
$$

$$
\begin{aligned}
c_{12} = &-\frac{n_1^2 \vartheta_{a-1}}{4x_2^4} \left\{ t_{13} \left[(4-3a)a^3 - \left(2a(a+2) - n_1^2 \right) n_1^2 \right] + (t_{14} + t_{32}) n_2^2 \left[a(a-2) + n_1^2 \right] \right. \\
&+ (4t_{14} + 4t_{32} + t_{24}) n_2^2 + (t_{23} + 5t_{13} - t_{14}) \left(2a^3 + 2an_1^2 - 3a^2 + n_1^2 \right) \\
&\left. + (4t_{14} - 4t_{23} - 7t_{13} + t_{24}) \left[a(a-2) + n_1^2 \right] - 3(t_{14} + t_{24} + t_{32}) n_2^2 - 3(t_{23} + t_{13} - t_{14} - t_{24}) \right\}
\end{aligned}
$$

$$
\begin{aligned}
c_{13} = &\ \frac{n_1 \vartheta_{a-0.5}}{8x_2^3} \left\{ t_{35} \left[(2a-1)a + 2n_1^2 \right] n_2^2 - t_{15} \left[(2a-1)a^3 + 3an_1^2 - 2n_1^4 \right] + \left(n_1^2 - a^2 + 2a^3 + 2an_1^2 \right) \right. \\
&\left. \times (2t_{15} + t_{25}) - 2t_{25} \left[(2a-1)a + 2n_1^2 \right] \right\} - \frac{n_1 \vartheta_{a-0.5}}{8x_2^3} \left\{ -F_3 \left[a(1+2a) + 2n_1^2 \right] x_2^2 + t_{35}(2a+1) n_2^2 \right\}
\end{aligned}
$$

$$
c_{14} = \frac{n_2 n_1^2 \vartheta_{a-0.5}}{8x_2^3} \left\{ 2(t_{38} + t_{18}) \left[(1-a)a - n_1^2 \right] - 2t_{18} + 3t_{28} - 2t_{38} \right\}
$$

$$\text{(A7)}$$

$$c_{21} = \frac{n_2^2 n_1^2 \vartheta_a}{4x_2^2}\left[t_{21}n_2^2 + (t_{22} - t_{31})\left(a^2 - 1 + n_1^2\right) - t_{31} + t_{22} - t_{21}\right]$$

$$c_{22} = \frac{n_1^2 n_2^2 \vartheta_{a-0.5}}{8x_2^3}\left\{2(t_{32} + t_{23})\left[(a-1)a + n_1^2\right] + 2t_{24}n_2^2 + t_{32} + t_{23} - t_{24}\right\}$$

$$c_{23} = n_1 n_2^2\left\{\frac{(t_{25}+t_{35})\left[(2a-1)a+2n_1^2\right]\vartheta_{a-0.5}}{8x_2^3} + \frac{at_{35}\vartheta_a}{4x_2^2}\right\}$$

$$c_{24} = -\frac{n_1^2 n_2}{4}\left\{\frac{\left[t_{38}\left(n_1^2+a^2\right)+t_{28}n_2^2\right]\vartheta_a}{x_2^2} + F_4\vartheta_{a+1}\right\}$$

$$\text{(A8)}$$

$$\begin{aligned}
c_{31} = {}& \frac{n_1^2 \vartheta_a}{4x_2^3}\Big\{q_{11}n_2^4 + (q_{31} + q_{21} + q_{12})n_2^2\left(a^2 - 1 + n_1^2\right) + (2q_{31} + 3q_{21} + q_{12})n_2^2 \\
& -(q_{31} + 2q_{21} + 2q_{11})n_2^2 + q_{22}\left[n_1^4 - (a+1)^3(3a-1) - 2(a+3)(a+1)n_1^2\right] \\
& +(4q_{22} + q_{12} - q_{21})\left(2n_1^2 a - 1 + 3a^2 + 3n_1^2 + 2a^3\right) \\
& -(5q_{22} + 3q_{12} - 3q_{21} - q_{11})\left(n_1^2 + a^2 - 1\right) + 2(q_{11} + q_{21} - q_{22} - q_{12})\Big\} \\
c_{32} = {}& -\frac{n_1^2 \vartheta_{a-0.5}}{8x_2^4}\Big\{-2q_{14}n_2^4 + 2(q_{32} - q_{13} - q_{24})\left(a^2 - a + n_1^2\right) - (q_{13} - 2q_{32} + 3q_{24})n_2^2 \\
& -2(q_{32} - 2q_{24} - 2q_{14})n_2^2 - 2q_{23}\left[(2 - 3a)a^3 - 2n_1^2 a(a+1) + n_1^4\right] \\
& -(q_{13} - q_{24} + 4q_{23})\left(4a^3 - 3a^2 + 4n_1^2 a + n_1^2\right) - 2(q_{14} - 3q_{13} + 3q_{24} - 5q_{23})\left(a^2 - a + n_1^2\right) \\
& +(q_{14} - 3q_{13} + 3q_{24} - 5q_{23})\Big\} + \frac{n_1^2(a^2+n_1^2)\vartheta_a \cot\gamma}{4x_2^3} \\
c_{33} = {}& \frac{(a^2+n_1^2)n_1\vartheta_a}{4x_2^3}\left[(q_{35} + q_{15})n_2^2 - q_{25}\left(a^2 - n_1^2\right) - (q_{25} + q_{15})a - q_{15}\right] \\
c_{34} = {}& \frac{n_1^2 \vartheta_a}{4x_2^3}\left[q_{18}n_2 - q_{18}n_2^3 - (q_{38} + q_{28})n_2\left(n_1^2 + a^2\right)\right]
\end{aligned}$$

$$\text{(A9)}$$

$$\begin{aligned}
c_{41} &= -\frac{\left(a^2+n_1^2\right)n_1^2\vartheta_a \cot\gamma}{4x_2^2} \\
c_{42} &= -P_1 c_{P1} - P_2 c_{P2} \\
c_{43} &= -\frac{n_1 \vartheta_{a+0.5}}{4x_2}\left\{F_3\left[(a + 0.5)a + n_1^2\right] + F_4(a + 0.5)\right\} \\
c_{44} &= \frac{F_4 n_1^2 n_2 \vartheta_{a+0.5}}{4x_2}
\end{aligned}$$

$$\text{(A10)}$$

with

$$\begin{aligned}
c_{P_1} &= -\frac{n_1^2(2n_1^2+2a^2+2a-1)\vartheta_{a+0.5}}{8x_2 \cot\gamma}, \\
c_{P_2} &= -\frac{\left(2n_2^2+1\right)n_1^2\vartheta_{a+0.5}}{8x_2 \cot\gamma} \\
c_{PH} &= c_{P_1} + c_{P_2} = -\frac{\left[2n_1^2+2a^2+2a+1+4n_2^2\right]n_1^2\vartheta_{a+0.5}}{16x_2 \cot\gamma}
\end{aligned}$$

$$\text{(A11)}$$

References

1. Shen, H.S.; Chen, T.Y. Buckling and postbuckling behaviour of cylindrical shells under combined external pressure and axial compression. *Thin-Walled Struct.* **1991**, *12*, 321–334. [CrossRef]
2. Winterstetter, T.A.; Schmidt, H. Stability of circular cylindrical steel shells under combined loading. *Thin-Walled Struct.* **2002**, *40*, 893–910. [CrossRef]
3. Tafresi, A.; Bailey, C.G. Instability of imperfect composite cylindrical shells under combined loading. *Compos. Struct.* **2007**, *80*, 49–64. [CrossRef]
4. Ajdari, M.A.B.; Jalili, S.; Jafari, M.; Zamani, J.; Shariyat, M. The analytical solution of the buckling of composite truncated conical shells under combined external pressure and axial compression. *J. Mech. Sci. Technol.* **2012**, *26*, 2783–2791. [CrossRef]
5. Sofiyev, A.H. On the buckling of composite conical shells resting on the Winkler–Pasternak elastic foundations under combined axial compression and external pressure. *Compos. Struct.* **2014**, *113*, 208–215. [CrossRef]
6. Shen, H.S.; Noda, N. Postbuckling of FGM cylindrical shells under combined axial and radial mechanical loads in thermal environments. *Int. J. Solid Struct.* **2005**, *42*, 4641–4662. [CrossRef]
7. Sofiyev, A.H. The buckling of FGM truncated conical shells subjected to combined axial tension and hydrostatic pressure. *Compos. Struct.* **2010**, *92*, 488–498. [CrossRef]

8. Sofiyev, A.H. Buckling analysis of FGM circular shells under combined loads and resting on the Pasternak type elastic foundation. *Mech. Res. Commun.* **2010**, *37*, 539–544. [CrossRef]

9. Khazaeinejad, P.; Najafizadeh, M.M.; Jenabi, J.; Isvandzibaei, M.R. On the buckling of functionally graded cylindrical shells under combined external pressure and axial compression. *J. Press. Vessel Technol.* **2010**, *132*, 064501. [CrossRef]

10. Huang, H.; Han, Q.; Feng, N.; Fan, X. Buckling of functionally graded cylindrical shells under combined loads. *Mech. Adv. Mater. Struct.* **2011**, *18*, 337–346. [CrossRef]

11. Wu, C.P.; Chen, Y.C.; Peng, S.T. Buckling analysis of functionally graded material circular hollow cylinders under combined axial compression and external pressure. *Thin-Walled Struct.* **2013**, *69*, 54–66. [CrossRef]

12. Shen, H.S.; Wang, H. Nonlinear bending and postbuckling of FGM cylindrical panels subjected to combined loadings and resting on elastic foundations in thermal environments. *Compos. B Eng.* **2015**, *78*, 202–213. [CrossRef]

13. Zhang, Y.; Huang, H.; Han, Q. Buckling of elastoplastic functionally graded cylindrical shells under combined compression and pressure. *Compos. B Eng.* **2015**, *69*, 120–126. [CrossRef]

14. Sofiyev, A.H.; Kuruoglu, N. The stability of FGM truncated conical shells under combined axial and external mechanical loads in the framework of the shear deformation theory. *Compos. B Eng.* **2016**, *92*, 463–476. [CrossRef]

15. Iijima, S. Synthesis of carbon nanotubes. *Nature* **1994**, *354*, 56–58. [CrossRef]

16. Kit, O.O.; Tallinen, T.; Mahadevan, L.; Timonen, J.; Koskinen, P. Twisting graphene nanoribbons into carbon nanotubes. *Phys. Rev. B* **2012**, *85*, 085428. [CrossRef]

17. Lim, H.E.; Miyata, Y.; Kitaura, R.; Nishimura, Y.; Nishimoto, Y.; Irle, S.; Warner, J.H.; Kataura, H.; Shinohara, H. Growth of carbon nanotubes via twisted graphene nanoribbons. *Nat. Commun.* **2013**, *4*, 2548. [CrossRef]

18. Bouaziz, O.; Kim, H.S.; Estrin, Y. Architecturing of metal-Based composites with concurrent nanostructuring: A new paradigm of materials design. *Adv. Eng. Mater.* **2013**, *15*, 336–340. [CrossRef]

19. Beygelzimer, Y.; Estrin, Y.; Kulagin, R. Synthesis of hybrid materials by severe plastic deformation: a new paradigm of SPD processing. *Adv. Eng. Mater.* **2015**, *17*, 1853–1861. [CrossRef]

20. Estrin, Y.; Beygelzimer, Y.; Kulagin, R. Design of architectured materials based on mechanically driven structural and compositional patterning. *Adv. Eng. Mater.* **2019**, *21*, 1900487. [CrossRef]

21. Ashrafi, B.; Hubert, P.; Vengallatore, S. Carbon nanotube-Reinforced composites as structural materials for micro actuators in micro electromechanical systems. *Nanotechnology* **2006**, *17*, 4895–4903. [CrossRef]

22. Rafiee, M.; Yang, J.; Kitipornchai, S. Thermal bifurcation buckling of piezoelectric carbon nanotube reinforced composite beams. *Comput. Math. Appl.* **2013**, *66*, 1147–1160. [CrossRef]

23. Araneo, R.; Bini, F.; Rinaldi, A.; Notargiacomo, A.; Pea, M.; Celozzi, S. Thermal-Electric model for piezoelectric ZnO nanowires. *Nanotechnology* **2015**, *26*, 265402. [CrossRef]

24. Ng, T.Y.; Lam, K.Y.; Liew, K.M. Effects of FGM materials on the parametric resonance of plate structures. *Comput. Meth. Appl. Mech. Eng.* **2000**, *190*, 953–962. [CrossRef]

25. Praveen, G.N.; Reddy, J.N. Nonlinear transient thermo elastic analysis of functionally graded ceramic–Metal plates. *Int. J. Solids Struct.* **1998**, *33*, 4457–4476. [CrossRef]

26. He, X.Q.; Ng, T.Y.; Sivashanker, S.; Liew, K.M. Active control of FGM plates with integrated piezoelectric sensors and actuators. *Int. J. Solids Struct.* **2001**, *38*, 1641–1655. [CrossRef]

27. Tornabene, F. Free vibration analysis of functionally graded conical, cylindrical shell and annular plate structures with a four-Parameter power-law distribution. *Comput. Meth. Appl. Mech. Eng.* **2009**, *198*, 2911–2935. [CrossRef]

28. Tornabene, F.; Viola, E.; Inman, D.J. 2-D differential quadrature solution for vibration analysis of functionally graded conical, cylindrical shell and annular plate structures. *J. Sound Vib.* **2009**, *328*, 259–290. [CrossRef]

29. Akbari, M.; Kiani, Y.; Aghdam, M.M.; Eslami, M.R. Free vibration of FGM Levy conical panels. *Compos. Struct.* **2015**, *116*, 732–746. [CrossRef]

30. Zhao, X.; Liew, K.M. Free vibration analysis of functionally graded conical shell panels by a meshless method. *Compos. Struct.* **2015**, *120*, 90–97. [CrossRef]

31. Jooybar, N.; Malekzadeh, P.; Fiouz, A.; Vaghefi, M. Thermal effect on free vibration of functionally graded truncated conical shell panels. *Thin-Walled Struct.* **2016**, *103*, 45–61. [CrossRef]

32. Kiani, Y.; Dimitri, R.; Tornabene, F. Free vibration study of composite conical panels reinforced with FG-CNTs. *Eng. Struct.* **2018**. [CrossRef]

33. Shen, H.S.; Xiang, Y. Postbuckling of nanotube-Reinforced composite cylindrical shells under combined axial and radial mechanical loads in thermal environment. *Compos. B Eng.* **2013**, *52*, 311–322. [CrossRef]
34. Sahmani, S.; Aghdam, M.M.; Bahrami, M. Nonlinear buckling and postbuckling behavior of cylindrical nano-Shells subjected to combined axial and radial compressions incorporating surface stress effects. *Compos. B Eng.* **2015**, *79*, 676–691. [CrossRef]
35. Heydarpour, Y.; Malakzadeh, P. Dynamic stability of rotating FG-CNTRC cylindrical shells under combined static and periodic axial loads. *Int. J. Struct. Stab. Dyn.* **2018**, *18*. [CrossRef]
36. Liew, K.M.; Lei, Z.X.; Zhang, L.W. Mechanical analysis of functionally graded carbon nanotube reinforced composites: A review. *Compos. Struct.* **2015**, *120*, 90–97. [CrossRef]
37. Jam, J.E.; Kiani, Y. Buckling of pressurized functionally graded carbon nanotube reinforced conical shells. *Compos. Struct.* **2015**, *125*, 586–595. [CrossRef]
38. Mirzaei, M.; Kiani, Y. Thermal buckling of temperature dependent FG-CNT reinforced composite conical shells. *Aerosp. Sci. Technol.* **2015**, *47*, 42–53. [CrossRef]
39. Ansari, R.; Torabi, J. Numerical study on the buckling and vibration of functionally graded carbon nanotube-Reinforced composite conical shells under axial loading. *Compos. B Eng.* **2016**, *95*, 196–208. [CrossRef]
40. Ansari, R.; Torabi, J.; Shojaei, M.F.; Hasrati, E. Buckling analysis of axially-Loaded functionally graded carbon nanotube-Reinforced composite conical panels using a novel numerical variational method. *Compos. Struct.* **2016**, *157*, 398–411. [CrossRef]
41. Mehri, M.; Asadi, H.; Wang, Q. Buckling and vibration analysis of a pressurized CNT reinforced functionally graded truncated conical shell under an axial compression using HDQ method. *Comput. Meth. Appl. Mech. Eng.* **2016**, *303*, 75–100. [CrossRef]
42. Fan, J.; Huang, J.; Ding, J. Free vibration of functionally graded carbon nanotube-Reinforced conical panels integrated with piezoelectric layers subjected to elastically restrained boundary conditions. *Adv. Mech. Eng.* **2017**, *9*, 1–17. [CrossRef]
43. Belal Ahmed Mohamed, M.S.; Zishun, L.; Kim, M.L. Active control of functionally graded carbon nanotube-Reinforced composite plates with piezoelectric layers subjected to impact loading. *J. Vib. Contr.* **2019**, *0*, 1–18.
44. Kiani, Y. Buckling of functionally graded graphene reinforced conical shells under external pressure in thermal environment. *Compos. B Eng.* **2019**, *156*, 128–137. [CrossRef]
45. Sofiyev, A.H.; Türkaslan-Esencan, B.; Bayramov, R.P.; Salamci, M.U. Analytical solution of stability of FG-CNTRC conical shells under external pressures. *Thin-Walled Struct.* **2019**, *144*. [CrossRef]
46. Ambartsumian, S.A. *Theory of Anisotropic Shells*; TT F-118; NASA: Washington, DC, USA, 1964.
47. Volmir, A.S. *Stability of Elastic Systems*; Nauka: Moscow, Russia, 1967.
48. Eslami, M.R. *Buckling and Postbuckling of Beams, Plates and Shells*; Springer: Berlin/Heidelberg, Germany, 2018.

Article

Buckling Behavior of Nanobeams Placed in Electromagnetic Field Using Shifted Chebyshev Polynomials-Based Rayleigh-Ritz Method

Subrat Kumar Jena [1], Snehashish Chakraverty [1] and Francesco Tornabene [2,*]

[1] Department of Mathematics, National Institute of Technology Rourkela, Rourkela 769008, India; sjena430@gmail.com (S.K.J.); sne_chak@yahoo.com (S.C.)

[2] Department of Innovation Engineering, University of Salento, 73100 Lecce, Italy

* Correspondence: francesco.tornabene@unisalento.it

Received: 17 July 2019; Accepted: 11 September 2019; Published: 16 September 2019

Abstract: In the present investigation, the buckling behavior of Euler–Bernoulli nanobeam, which is placed in an electro-magnetic field, is investigated in the framework of Eringen's nonlocal theory. Critical buckling load for all the classical boundary conditions such as "Pined–Pined (P-P), Clamped–Pined (C-P), Clamped–Clamped (C-C), and Clamped-Free (C-F)" are obtained using shifted Chebyshev polynomials-based Rayleigh-Ritz method. The main advantage of the shifted Chebyshev polynomials is that it does not make the system ill-conditioning with the higher number of terms in the approximation due to the orthogonality of the functions. Validation and convergence studies of the model have been carried out for different cases. Also, a closed-form solution has been obtained for the "Pined–Pined (P-P)" boundary condition using Navier's technique, and the numerical results obtained for the "Pined–Pined (P-P)" boundary condition are validated with a closed-form solution. Further, the effects of various scaling parameters on the critical buckling load have been explored, and new results are presented as Figures and Tables. Finally, buckling mode shapes are also plotted to show the sensitiveness of the critical buckling load.

Keywords: buckling; electromagnetic field; nanobeam; shifted chebyshev polynomial; rayleigh-ritz method

---

1. Introduction

In recent decades, with the advent of technological advancement, small-scale structures like nanoclock, nano-oscillator, nanosensor, NEMS (Nano-Electro-Mechanica System), and so forth have found tremendous attention due to their various applications. In this scenario, the study of dynamical behaviors of nanostructures is important and a need of the hour. Due to the small size of nanostructures, experimental analysis of such structures is always challenging and difficult as it requires enormous experimental efforts. Moreover, classical mechanics or continuum mechanics fail to address the nonlocal effect also. In this regard, nonlocal continuum theories were recently found to be useful for the modeling of micro- and nanosized structures. Various researchers developed different nonlocal or nonclassical continuum theories to assimilate the small-scale effect, such as strain gradient theory [1], couple stress theory [2], micropolar theory [3], nonlocal elasticity theory [4], and so on. Out of all these theories, nonlocal elasticity theory of Eringen has been broadly used for the dynamic analysis of nanostructures. Few studies regarding the vibration and buckling of beam, membrane, and nanostructures such as nanobeam, nanotube, nanoribbon, and so forth can be found in [5–14].

Wang et al. [15] studied buckling behavior of micro- and nanorods/tubes with the help of Timoshenko beam theory, where small-scale effect was addressed by using the nonlocal elasticity theory of Eringen. Emam [16] incorporated nonlocal elasticity theory to analyze the buckling and the postbuckling response of nanobeams analytically. Yu et al. [17] used nonlocal thermo-elasticity theory to study buckling behavior of Euler–Bernoulli nanobeam with nonuniform temperature distribution. Nejad et al. [18] employed a generalized differential quadrature method to undertake the buckling analysis of the nanobeams made of two-directional functionally graded materials (FGM) using nonlocal elasticity theory. Dai et al. [19] analytically studied the prebuckling and postbuckling behavior of nonlocal nanobeams subjected to the axial and longitudinal magnetic forces. Bakhshi Khaniki and Hosseini-Hashemi [20] implemented nonlocal strain gradient theory to investigate the buckling behavior of Euler–Bernoulli beam, considering different types of cross-section variation using the generalized differential quadrature method. Yu et al. [21] proposed a three characteristic-lengths-featured size-dependent gradient-beam model by incorporating the modified nonlocal theory and Euler–Bernoulli beam theory. He implemented the weighted residual approach to solve the six-order differential equation to investigate the buckling behaviors. Malikan [22] used a refined beam theory to investigate the buckling behavior of SWCNT (Single Walled carbon NanoTube) using Navier's method. Here, unidirectional load is applied on the SWCNT. Buckling analysis of FG (Fanctionally Graded) nanobeam was studied in [23] analytically with the help of Navier's method under the framework of first-order shear deformation beam theory. Malikan et al. investigated the transient response [24] of nanotube for a simply supported boundary condition using Kelvin–Voigt viscoelasticity model with nonlocal strain gradient theory. An investigation regarding damped forced vibration of SWCNTs using a shear deformation beam theory subjected to viscoelastic foundation and thermal environment can be found in [25]. Some other notable studies can be seen in [26–30]

As per the present authors' knowledge, the buckling behavior of the Euler–Bernoulli nanobeam placed in an electro-magnetic field using shifted Chebyshev polynomials Rayleigh-Ritz method has been studied for the first time for "Pined–Pined (P-P), Clamped–Pined (C-P), Clamped–Clamped (C-C), and Clamped-Free (C-F)" boundary conditions. Euler–Bernoulli nanobeam is combined with Hamilton's principle to derive the governing equation. Also, a closed-form solution for the Pined–Pined (P-P) boundary condition has been obtained by using the Navier's technique. Critical buckling load for all the classical boundary conditions were obtained and a parametric study has been carried out to comprehend the effects of various scaling parameters on the critical buckling load through graphical and tabular results. Further, buckling mode shapes for different boundary conditions were drawn to show the sensitivity towards various scaling parameters.

2. Proposed Model for Electromagnetic Nanobeam

In this study, the nanobeam with length L and diameter d is placed in an electromagnetic field with the electric field intensity as E and the magnetic flux density as B. The schematic diagram for continuum model of the nanobeam is shown in Figure 1. Then, by Ohm's law, the current density (J) of the system due to the induced current (because of Lorentz force) is given as [31]

$$J = \sigma_0(E + w_0 \times B) = \sigma_0(E + w_0 \times \mu_0 H) \tag{1}$$

where σ_0 is the electrical conductivity, μ_0 is the magnetic permeability of free space, and H is the magnetic field strength. By neglecting the electric field intensity, the nanobeam experiences a magnetic force or Pondermotive force which is denoted by f_{em} and can be expressed as [31]

$$f_{em} = J \times B = \sigma_0(E + w_0 \times \mu_0 H) \times \mu_0 H = \sigma_0 \mu_0^2 H^2 w_0 \tag{2}$$

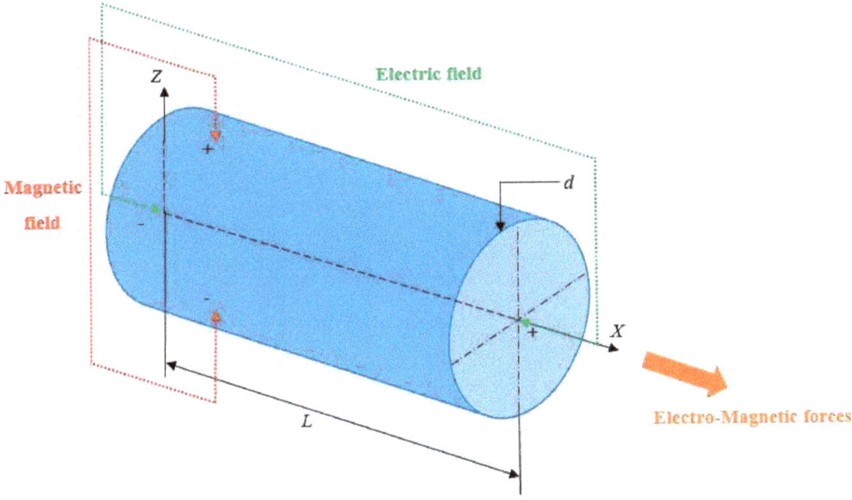

Figure 1. Schematic continuum model of the nanobeam placed in electromagnetic field.

According to Euler–Bernoulli beam theory, the displacement field at any point may be stated as [32]

$$u_1(x,\, z,\, t) = -z\frac{\partial w_0\,(x,\,t)}{\partial x} \tag{3a}$$

$$u_3(x,\, z,\, t) = w_0\,(x,\,t) \tag{3b}$$

Here, u_1 and u_3 represent displacements along x and z directions, respectively, and $w_0\,(x,\,t)$ denotes the transverse displacement of the point on the mid-plane of the beam. The strain-displacement relation may be expressed as

$$\varepsilon_{xx} = -z\frac{\partial^2 w_0\,(x,\,t)}{\partial x^2} \tag{4}$$

Under the framework of Euler–Bernoulli nanobeam, the variation of strain energy (δU) and the variation of work done by external force (δW_e) are presented as

$$\delta U = \int_0^L \int_A \sigma_{xx}\delta\varepsilon_{xx}dA\,dx = \int_0^L \left[-M_{xx}\frac{\partial^2 \delta w_0}{\partial x^2}\right]dx, \tag{5}$$

$$\delta W_e = \int_0^L \left[P\left(\frac{dw_0}{dx}\right)\left(\frac{d\delta w_0}{dx}\right) + \sigma_0\mu_0^2 H^2 w_0\delta w_0\right]dx, \tag{6}$$

where σ_{xx} is the normal stress, ε_{xx} is the normal strain, and $M_{xx} = \int_A \sigma_{xx}\,z\,dA$ is the bending moment of nanobeam. The Hamilton's principle for the conservative system is stated as

$$\delta\prod = \int_0^t \delta\,(W_e + U)\,dt, \tag{7}$$

Substituting Equations (5)–(7) and setting $\delta \Pi = 0$, we have

$$
\begin{aligned}
\delta \Pi &= \int_0^t \int_0^L \left[P\left(\frac{dw_0}{dx}\right)\left(\frac{d\delta w_0}{dx}\right) + \sigma_0 \mu_0^2 H^2 w_0 \delta w_0 - M_{xx}\frac{\partial^2 \delta w_0}{\partial x^2} \right] dxdt \\
&= \int_0^t \int_0^L \left[-P\left(\frac{d^2 w_0}{dx^2}\right)\delta w_0 + \sigma_0 \mu_0^2 H^2 w_0 \delta w_0 + \frac{\partial^2 M_{xx}}{\partial x^2}\delta w_0 \right] dxdt
\end{aligned}
\tag{8}
$$

The equation of motion for buckling behavior can be obtained as

$$
\frac{d^2 M_{xx}}{dx^2} + \sigma_0 \mu_0^2 H^2 w_0 = P\frac{d^2 w_0}{dx^2}
\tag{9}
$$

For an isotropic nonlocal beam, the nonlocal elasticity theory of Eringen can be expressed as [4]

$$
\left(1 - \mu\frac{\partial^2}{\partial x^2}\right)\sigma_{xx} = E\varepsilon_{xx}
\tag{10}
$$

where $\mu = (e_0 a)^2$ is the nonlocal parameter, E is Young's modulus. Here e_0 and a denote material constant and internal characteristic length, respectively. Multiplying Equation (10) by zdA and integrating over A, the nonlocal constitutive relation for Euler–Bernoulli nanobeam may be expressed as

$$
M_{xx} - \mu\frac{d^2 M_{xx}}{dx^2} = -EI\frac{d^2 w_0}{dx^2}
\tag{11}
$$

where $I = \int_A z^2 dA$, is the second moment of area. Using Equation (9) in Equation (11) and rearranging, the nonlocal bending moment can be obtained as

$$
M_{xx} = -EI\frac{d^2 w_0}{dx^2} + \mu P\frac{d^2 w_0}{dx^2} - \mu\sigma_0 \mu_0^2 H^2 w_0
\tag{12}
$$

Equating the nonlocal strain energy with work done by an external force, we obtain

$$
-\int_0^L \left(-EI\frac{d^2 w_0}{dx^2} + \mu P\frac{d^2 w_0}{dx^2} - \mu\sigma_0 \mu_0^2 H^2 w_0\right)\frac{d^2 w_0}{dx^2}\,dx = \int_0^L \left[P\left(\frac{dw_0}{dx}\right)^2 + \sigma_0 \mu_0^2 H^2 w_0^2\right]dx
\tag{13}
$$

Substituting Equation (12) in Equation (9), we obtain the governing equation of motion as

$$
-EI\frac{d^4 w_0}{dx^4} + \mu P\frac{d^4 w_0}{dx^4} - \mu\sigma_0 \mu_0^2 H^2\frac{d^2 w_0}{dx^2} + \sigma_0 \mu_0^2 H^2 w_0 = p\frac{d^2 w_0}{dx^2}
\tag{14}
$$

Let us define the following nondimensional parameters

$X = \frac{x}{L}$ = nondimensional spatial coordinate

$W = \frac{w_0}{L}$ = nondimensional transverse displacement

$\hat{P} = \frac{PL^2}{EI}$ = dimensionless frequency parameter

$\alpha = \frac{e_0 a}{L}$ = dimensionless nonlocal parameter

$H_a^2 = \frac{\sigma_0 \mu_0^2 H^2 L^4}{EI}$ = dimensionless Hartmann parameter.

Incorporating the above nondimensional parameters in Equations (13) and (14), we have

$$
\int_0^1 \left\{ \left(\frac{d^2 W}{dX^2}\right)^2 + \alpha^2 H_a^2\left(W\frac{d^2 W}{dX^2}\right) - H_a^2 W^2 \right\} dX = \hat{P}\int_0^1 \left\{ \left(\frac{dW}{dX}\right)^2 + \alpha^2\left(\frac{d^2 W}{dX^2}\right)^2 \right\} dX
\tag{15}
$$

$$\frac{d^4W}{dX^4} + \alpha^2 H_a^2 \frac{d^2W}{dX^2} - H_a^2 W = \hat{P}\left(\alpha^2 \frac{d^4W}{dX^4} - \frac{d^2W}{dX^2}\right) \tag{16}$$

3. Solution Methodology

3.1. Shifted Chebyshev Polynomials-Based Rayleigh-Ritz Method

Chebyshev polynomials of the first kind $(C_n(X))$ are a sequence of orthogonal polynomials with $X \in [-1\ 1]$, and few terms of the sequence are defined as

$$C_0(X) = 1$$

$$C_1(X) = X \tag{17}$$

$$C_n(X) = 2XC_{n-1}(X) - C_{n-2}(X), \qquad n = 2, 3, \ldots$$

In order to solve Equation (12), Rayleigh-Ritz method is implemented along with Chebyshev polynomials of the first kind as shape function. For more details about the Rayleigh-Ritz method, one may refer to the books [33,34]. The main advantages of using Chebyshev polynomials over algebraic polynomials $\left(1,\ X,\ X^2,\ X^3,\ \ldots X^n\right)$ are the orthogonal properties of Chebyshev polynomials, which reduce the computational effort, and for the higher value of n $(n > 10)$, the system avoids ill-conditioning. Since the domain of the nanobeam lies in $[0\ 1]$, the Chebyshev polynomials must be reduced to shifted Chebyshev polynomials of the first kind $(C_n^*(X))$ with $X \in [0\ 1]$. This is achieved by transforming $X \mapsto 2X - 1$, and there exists a one-to-one correspondence between $[0\ 1]$ and $[-1\ 1]$. Accordingly, the first few terms of shifted Chebyshev polynomials of the first kind $(C_n^*(X))$ can be written as, (where $C_n^*(X) = C_n(2X - 1)$)

$$C_0^*(X) = 1$$

$$C_1^*(X) = 2X - 1 \tag{18}$$

$$C_n^*(X) = 2(2X - 1)C_{n-1}^*(X) - C_{n-2}^*(X), \qquad n = 2, 3, \ldots$$

The transverse displacement function $(W(X))$ as per the Rayleigh-Ritz method can now be expressed as

$$W(X) = X^p(1 - X)^q \sum_{i=1}^{N} a_i C_{i-1}^*(X) \tag{19}$$

where $a_i's$ are unknowns, C_{i-1}^* are the shifted Chebyshev polynomials of the index $i - 1$, N is the number of terms required to obtain the result with the anticipated accuracy, p and q are the exponents which decide the boundary conditions, as given in Table 1.

Table 1. Values of p and q for different boundary conditions [33,34].

Boundary Conditions	p	q
P-P	1	1
C-P	2	1
C-C	2	2
C-F	2	0

Replacing Equation (19) into Equation (15), and minimizing the buckling load parameter with respect to the coefficients of the admissible functions (i.e., $a_i's$, $i = 1, 2, 3 \ldots N$), we obtain the generalized eigenvalue problem as

$$[K]\{A\} = \hat{P}[B]\{A\} \tag{20}$$

where $A = [a_1\, a_2\, a_3\, \ldots a_N]^T$, $[K]$ is the stiffness matrix and $[B]$ is the buckling matrix, which are presented as

$$K(i,\, j) = \int_0^1 \left(2C_i^{*\prime\prime} C_j^{*\prime\prime} + \alpha^2 H_a^2 C_i^{*\prime\prime} C_j^* + \alpha^2 H_a^2 C_i^* C_j^{*\prime\prime} - 2H_a^2 C_i^* C_j^*\right) dX, \; i,\, j = 1,\, 2,\, 3, \ldots N$$

$$B(i,\, j) = \int_0^1 \left(2C_i^{*\prime} C_j^{*\prime} + 2\alpha^2 C_i^{*\prime\prime} C_j^{*\prime\prime}\right) dX, \; i,\, j = 1,\, 2,\, 3, \ldots N.$$

3.2. Closed-Form Solution for P-P Boundary Condition Using Navier's Technique

Navier's technique has been employed to find a closed-form solution for the Pined–Pined (P-P) boundary condition. As per the Navier's technique, the transverse displacement (W) may be expressed as [23–25]

$$W = \sum_{n=1}^{\infty} W_n \, \sin(n\pi X) \, e^{i\omega_n t} \tag{21}$$

In which W_n, and ω_n are the displacement and frequency of the beam. Now, by substituting Equation (21) in Equation (16), the buckling load $\hat{P}$ for Pined–Pined (P-P) boundary condition is calculated as

$$\hat{P}_n = \frac{(n\pi)^4 - \alpha^2 H_a^2 (n\pi)^2 - H_a^2}{\alpha^2 (n\pi)^4 + (n\pi)^2} \tag{22}$$

4. Numerical Results and Discussion

Shifted Chebyshev polynomials-based Rayleigh-Ritz method has been employed to convert Equation (15) into the generalized eigenvalue problem given in Equation (20). MATLAB codes have been utilized to solve the generalized eigenvalue problem and to compute the critical buckling load parameter. Likewise, Navier's technique has been adopted to find a closed-form solution for the P-P boundary condition, which is demonstrated in Equation (22). In this regard, the following parameters are taken from [15] for computation purpose

$$E = 1 \text{ Tpa}, \; d = 1 \text{ nm, and } L = 10 \text{ nm}.$$

4.1. Validation

Results of the present model were authenticated by two ways, firstly, by matching with the numerical results given by Wang et al. (2006) for "P-P, C-P, C-C, and C-F" boundary conditions and secondly, Pined–Pined results were compared with the closed-form solution obtained by Navier's technique. For this purpose, the Hartmann parameter (H_a) in the present model was taken as zero, and the critical buckling load parameters (P_{cr}) for "P-P, C-P, C-C, and C-F" boundary conditions were taken into investigation. Comparisons of critical buckling load (P_{cr}) are presented in Table 2. Similarly, Table 3 illustrates the comparison of the P-P boundary condition with the Navier's results, with $E = 1$ TPa, $d = 1$ nm, $L = 10$ nm, and $H_a = 1$. From these comparisons, it is evident that the critical buckling loads of the present model are on a par with [15] in the particular case and with Navier's solution for the P-P boundary condition.

Table 2. Comparison of "Critical buckling load" (P_{cr}) in nN with [15] for $\frac{L}{d} = 10$.

(a) Comparison of P-P and C-P boundary conditions

e_0a	P-P		C-P	
	Present	[15]	Present	[15]
0	4.8447	4.8447	9.9155	9.9155
0.5	4.7281	4.7281	9.4349	9.4349
1	4.4095	4.4095	8.2461	8.2461
1.5	3.9644	3.9644	6.8151	6.8151
2	3.4735	3.4735	5.4830	5.4830

(b) Comparison of C-C and C-F boundary conditions

e_0a	C-C		C-F	
	Present	[15]	Present	[15]
0	19.3790	19.3790	1.2112	1.2112
0.5	17.6381	17.6381	1.2037	1.2037
1	13.8939	13.8939	1.1820	1.1820
1.5	10.2630	10.2630	1.1475	1.1475
2	7.5137	7.5137	1.1024	1.1024

Table 3. Comparison of "Critical buckling load" (P_{cr}) in nN with Navier's closed-form solution for P-P boundary condition.

e_0a	Present (R-R)	Navier's Solution
0	4.7950	4.7950
0.5	4.6783	4.6783
1	4.3598	4.3598
1.5	3.9146	3.9146
2	3.4237	3.4237
2.5	2.9467	2.9467
3	2.5160	2.5160
3.5	2.1434	2.1434
4	1.8287	1.8287

4.2. Convergence

A convergence study has been performed to know the number of terms needed to obtain the results of critical buckling load parameters (P_{cr}) and verify the present model using the Rayleigh-Ritz method. In this regard, $H_a = 1$, $L = 10$, and $e_0a = 1$ were taken for computation purpose. Both the tabular and graphical results were noted for "P-P, C-P, C-C, and C-F" boundary conditions, which are demonstrated in Table 4 and Figure 2, respectively. The C-F boundary condition is converging faster with $N = 5$, whereas other edges such as P-P, C-P, and C-C take $N = 7$ for acquiring the desired accuracy. These results revealed that both the model and the results are useful regarding the present investigation.

Table 4. Effect of no. of terms (N) on critical buckling load (P_{cr}) with $H_a = 1$, $L = 10$, and $e_0a = 1$.

N	P-P	C-P	C-C	C-F
2	5.211151	8.452577	14.486540	1.100703
3	4.362030	8.240579	13.869377	1.094818
4	4.362029	8.209547	13.869177	1.094614
5	4.359792	8.208410	13.863597	1.094613
6	4.359790	8.208342	13.863594	1.094613
7	4.359791	8.208341	13.863584	1.094613
8	4.359791	8.208341	13.863584	1.094613
9	4.359791	8.208341	13.863584	1.094613
10	4.359791	8.208341	13.863584	1.094613

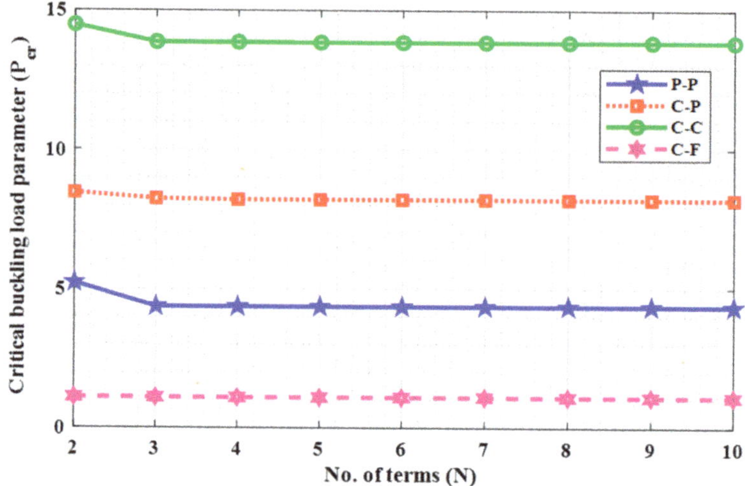

Figure 2. No. of terms (N) vs. critical buckling load (P_{cr}) with $H_a = 1$, $L = 10$, and $e_0 a = 1$.

4.3. Influence of Small Scale Parameter

This subsection is dedicated to investigating the influence of a small scale parameter $(e_0 a)$ on critical buckling load parameters and the critical buckling load ratio. The four frequently used boundary conditions such as "P-P, C-P, C-C, and C-F" were taken into consideration with $N = 7$, $L = 10$, and $H_a = 2$. Tabular and graphical results are illustrated in Table 4 and Figures 2 and 3 for different $e_0 a$ (0, 0.5, 1, 1.5, 2, 2.5, 3, 3.5, 4, 4.5, 5). Table 5 and Figure 3 represent the variation of small scale parameter $(e_0 a)$ on critical buckling load, whereas Figure 4 demonstrates the variation of small scale parameter $(e_0 a)$ on the critical buckling load ratio. The critical buckling load ratio may be defined as the ratio of critical buckling load calculated using nonlocal theory and classical theory $(e_0 a = 0)$. This critical buckling load ratio acts as an index to estimate the influence of the small scale parameter $(e_0 a)$ qualitatively on buckling load. From Table 5 and Figure 3, it is observed that critical buckling load is decreasing with an increase in small scale parameter $(e_0 a)$, and this decline is more in case of the C-C boundary condition. From Figure 4, it may also be noted that the influence of the small scale parameter is comparatively more in C-C edge and less in C-F edge.

Table 5. Effect of small scale parameter $(e_0 a)$ on critical buckling load (P_{cr}) in nN with $N = 7$, $L = 10$, and $H_a = 2$.

$e_0 a$	P-P	C-P	C-C	C-F
0	4.645787	9.748924	19.229661	0.840199
0.5	4.529126	9.275797	17.497785	0.836210
1	4.210584	8.094859	13.772744	0.824504
1.5	3.765433	6.673148	10.160354	0.805815
2	3.274518	5.349719	7.425368	0.781223
2.5	2.797456	4.255716	5.510414	0.751973
3	2.366762	3.397759	4.184610	0.719306
3.5	1.994208	2.737457	3.253694	0.684313
4	1.679487	2.230108	2.585265	0.647847
4.5	1.416723	1.837632	2.093720	0.610487
5	1.198278	1.530786	1.723927	0.572534

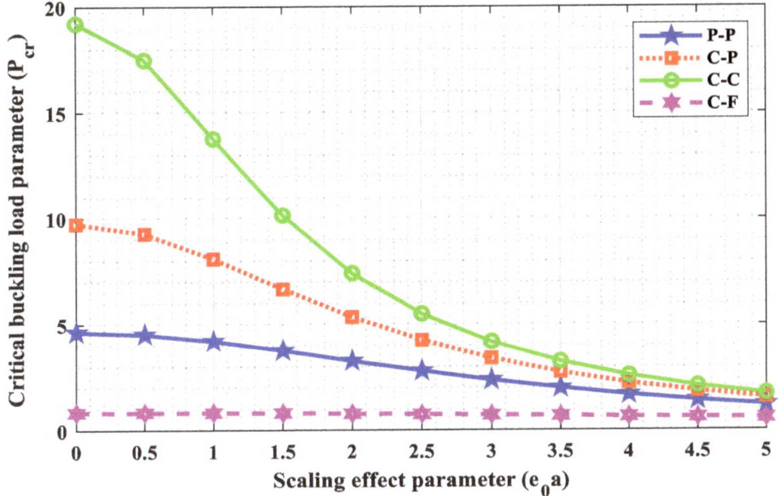

Figure 3. Variation of $(e_0 a)$ with (P_{cr}) for $N = 7$, $L = 10$, and $H_a = 2$.

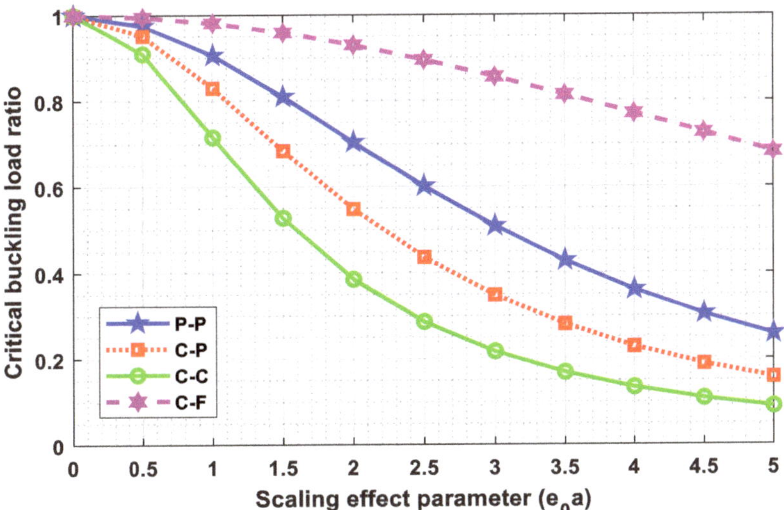

Figure 4. Small scale parameter $(e_0 a)$ vs. critical buckling load ratio.

4.4. Influence of Aspect Ratio

The objective of this subsection is to study the impact of aspect ratio (L/d) on the critical buckling load (P_{cr}) with "P-P, C-P, C-C, and C-F" boundary conditions for different L/d (5, 10, 15, 20, 25, 30, 35, 40, 45, 50). The effect of aspect ratio has been reported in Table 6 and Figure 5 for $N = 7$, $e_0 a = 1$, and $H_a = 2$, which are respectively. From this study, it is essential to note that the critical buckling load decreases with an increase in aspect ratio (L/d). This decrease is more consequential for the lower value of aspect ratio.

Table 6. Effect of aspect ratio (L/d) on critical buckling load (P_{cr}) in nN with $N = 7$, $e_0 a = 1$, and $H_a = 2$.

L/d	P-P	C-P	C-C	C-F
5	13.098075	21.398879	29.701473	3.124893
10	4.210584	8.094859	13.772744	0.824504
15	1.974312	3.972494	7.267483	0.370283
20	1.132281	2.318949	4.374446	0.209052
25	0.731275	1.510526	2.893474	0.134022
30	0.510359	1.059208	2.046611	0.093157
35	0.376087	0.782797	1.520631	0.068481
40	0.288505	0.601638	1.172838	0.052449
45	0.228261	0.476628	0.931406	0.041452
50	0.185069	0.386801	0.757197	0.033582

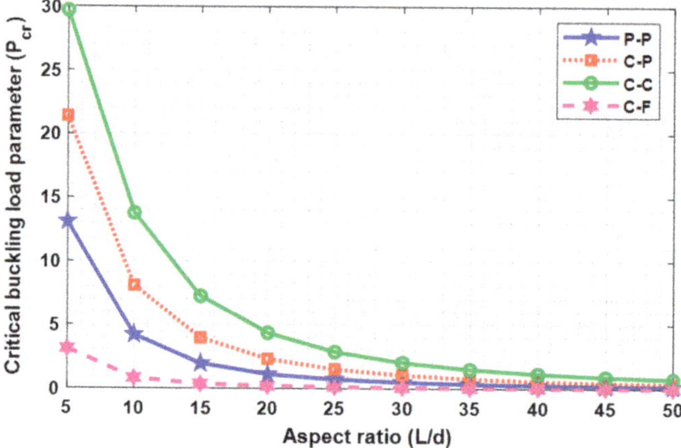

Figure 5. Variation of (L/d) with (P_{cr}) for $N = 7$, $e_0 a = 1$, and $H_a = 2$.

4.5. Influence of Hartmann Parameter

For the designing of electromagnetic devices, proper knowledge about the effect of electric and magnetic fields on critical buckling load is necessary as it greatly influences the lifespan of electromagnetic devices. In this regard, the effect of Hartmann parameter (H_a) on the critical buckling load (P_{cr}) has been studied in this subsection for different values of H_a (0, 1, 1.5, 2, 2.5, 3, 3.5). Table 7 and Figure 6 represent the results for the variation of critical buckling load with Hartmann parameter (H_a) for "P-P, C-P, C-C, and C-F" boundary conditions. From these results, we may note that the critical buckling load decreases with increase in Hartmann parameter, but this drop in critical buckling load is very slow.

Table 7. Effect of Hartmann parameter (H_a) on critical buckling load (P_{cr}) in nN with $N = 7$, $e_0 a = 1$, and $L = 10$.

H_a	P-P	C-P	C-C	C-F
0	4.409527	8.246144	13.893850	1.182017
0.5	4.397093	8.236694	13.886284	1.160290
1	4.359791	8.208341	13.863584	1.094613
1.5	4.297621	8.161070	13.825741	0.983502
2	4.210584	8.094859	13.772744	0.824504
2.5	4.098678	8.009677	13.704575	0.614226
3	3.961904	7.905484	13.621213	0.348424
3.5	3.800262	7.782236	13.522631	0.022134

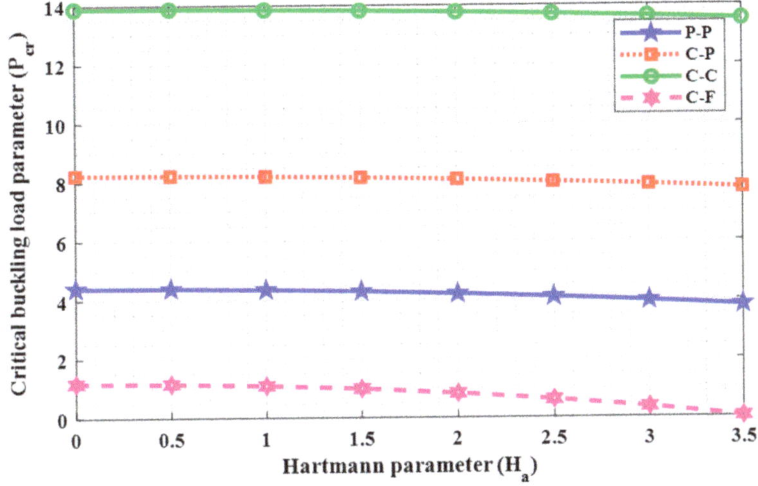

Figure 6. Response of (H_a) on (P_{cr}) for $N = 7$, $e_0 a = 1$, and $L = 10$.

5. Buckling Mode Shape

Buckling is the state of instability of structures that leads to structural failure. In this circumstance, the buckling mode shape has a vital role in predicting the instability. In this regard, buckling mode shapes were plotted with $H_a = 0.5$ and $L = 10$ for different $e_0 a$ (0.5, 1, 1.5, 2). Figures 7–10 show the buckling mode shapes for "P-P, C-P, C-C, and C-F" boundary conditions, respectively. From these figures, one may witness the sensitiveness of buckling mode shapes towards scaling parameters. Also, these mode shapes help to predict the mechanical health and lifespan of several electromechanical devices.

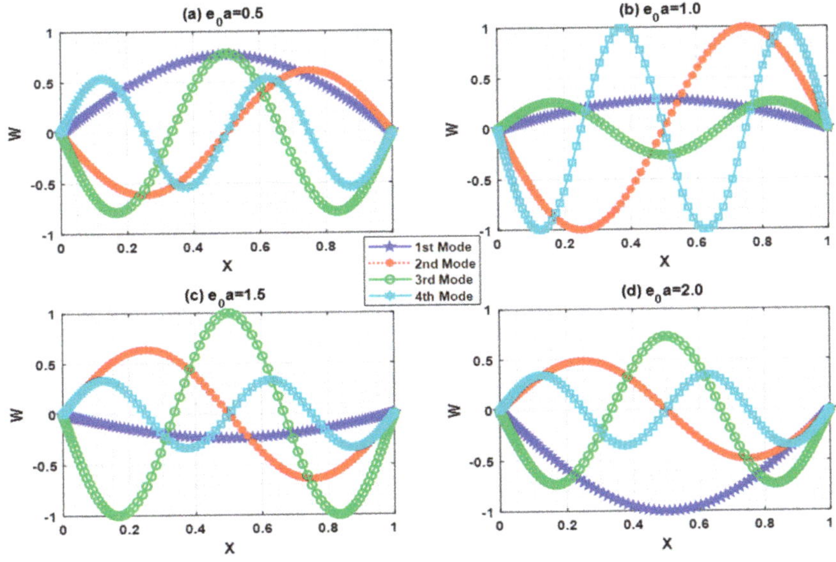

Figure 7. Buckling mode shape for P-P boundary condition with $H_a = 0.5$ and $L = 10$.

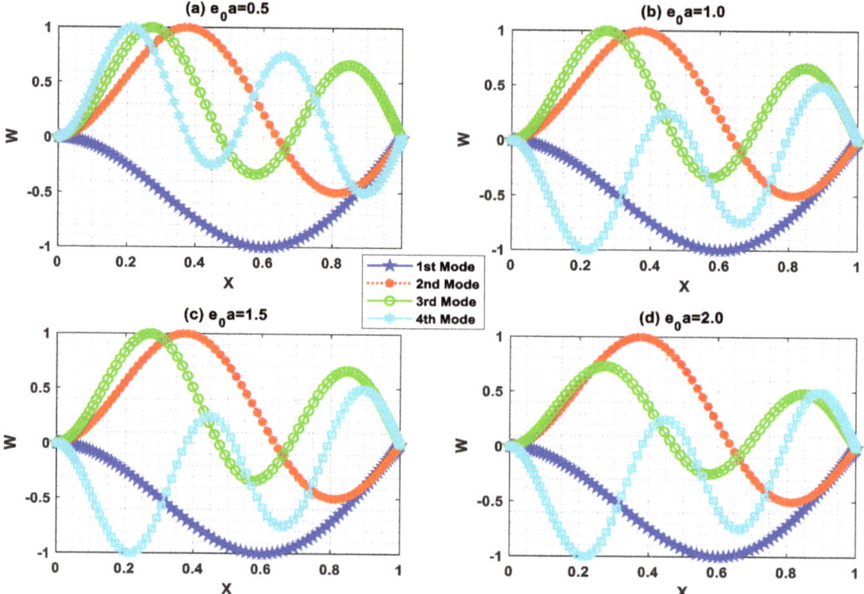

Figure 8. Buckling mode shape for C-P boundary condition with $H_a = 0.5$ and $L = 10$.

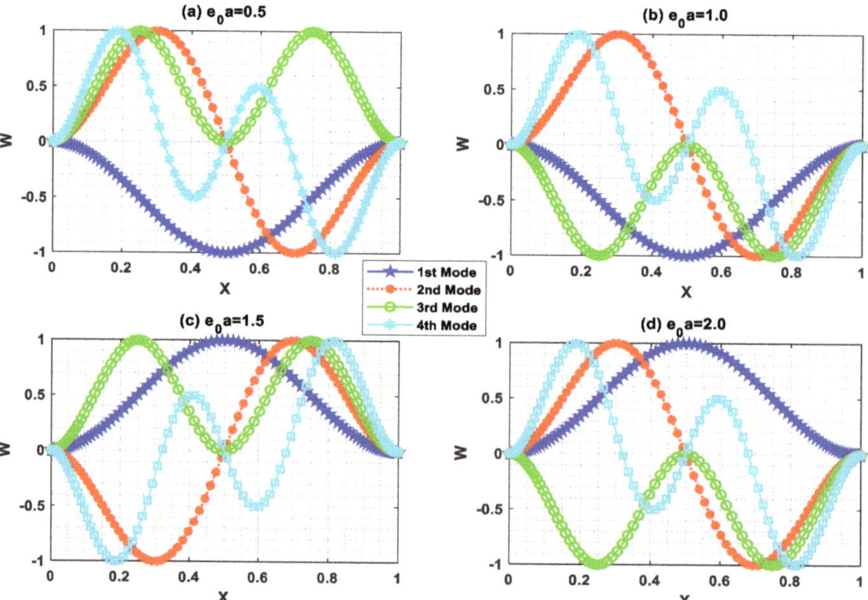

Figure 9. Buckling mode shape for C-C boundary condition with $H_a = 0.5$ and $L = 10$.

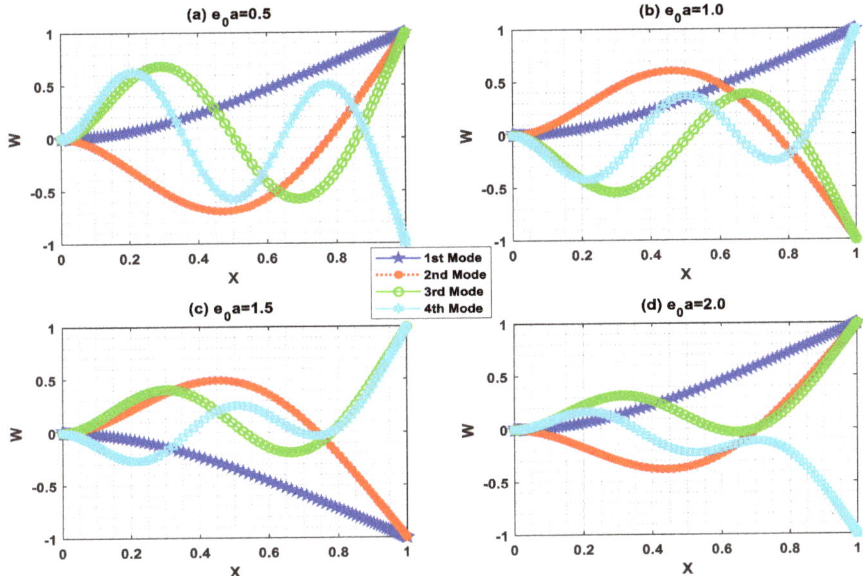

Figure 10. Buckling mode shape for C-F boundary condition with $H_a = 0.5$ and $L = 10$.

6. Concluding Remarks

The buckling behavior of Electromagnetic nanobeam is investigated in the combined framework of Euler–Bernoulli beam theory and Eringen's nonlocal theory. Critical buckling load parameters were obtained using shifted Chebyshev polynomials-based Rayleigh-Ritz method for all the classical boundary conditions such as "Pined–Pined (P-P), Clamped–Pined (C-P), Clamped–Clamped (C-C), and Clamped-Free (C-F)". The C-F boundary condition converges faster with $N = 5$, whereas other boundary conditions such as P-P, C-P, and C-C require $N = 7$ for achieving convergence to the desired accuracy. Critical buckling load parameters decrease with an increase in small scale parameter, and this decline is more in case of the C-C boundary condition. It may also be noted that the influence of small scale parameter is comparatively more in C-C edge and less in C-F edge. It is interesting to note that the critical buckling load decreases with an increase in aspect ratio. This decrease is more consequential for the lower values of aspect ratio. We may note that the critical buckling load decreases with an increase in Hartmann parameter, but this drop in critical buckling load is prolonged. The C-C nanobeam possesses the highest critical buckling load, whereas the C-F nanobeam possesses the lowest.

Author Contributions: Conceptualization, S.C., S.K.J. and F.T.; Formal analysis, S.C., S.K.J. and F.T.; Investigation, S.C., S.K.J. and F.T.; Validation, S.C., S.K.J. and F.T.; Writing-original draft, S.C., S.K.J. and F.T.; Writing-review & editing, S.C., S.K.J. and F.T.

Funding: This research received no external funding.

Acknowledgments: The first two authors are thankful to Defence Research & Development Organization (DRDO), Ministry of Defence, New Delhi, India (Sanction Code: DG/TM/ERIPR/GIA/17-18/0129/020) for the funding to carry out the present research work.

Conflicts of Interest: The authors declare no conflict of interest.

References

1. Aifantis, E.C. Strain gradient interpretation of size effects. *Int. J. Fract.* **1999**, *95*, 299–314. [CrossRef]
2. Mindlin, R.D.; Tiersten, H.F. Effects of couple-stresses in linear elasticity. *Arch. Ration. Mech. Anal.* **1962**, *11*, 415–448. [CrossRef]

3. Eringen, A.C. Theory of micropolar plates. *J. Appl. Math. Phys.* **1967**, *18*, 12–30. [CrossRef]
4. Eringen, A.C. Nonlocal polar elastic continua. *Int. J. Eng. Sci.* **1972**, *10*, 1–16. [CrossRef]
5. Chakraverty, S.; Jena, S.K. Free Vibration of Single Walled Carbon Nanotube Resting on Exponentially Varying Elastic Foundation. *Curved Layer. Struct.* **2018**, *5*, 260–272. [CrossRef]
6. Jena, S.K.; Chakraverty, S. Free vibration analysis of Euler–Bernoulli nanobeam using differential transform method. *Int. J. Comput. Mater. Sci. Eng.* **2018**, *7*, 1850020. [CrossRef]
7. Jena, S.K.; Chakraverty, S. Free Vibration Analysis of Variable Cross-Section Single Layered Graphene Nano-Ribbons (SLGNRs) Using Differential Quadrature Method. *Front. Built Environ.* **2018**, *4*, 63. [CrossRef]
8. Jena, S.K.; Chakraverty, S. Free Vibration Analysis of Single Walled Carbon Nanotube with Exponentially Varying Stiffness. *Curved Layer. Struct.* **2018**, *5*, 201–212. [CrossRef]
9. Jena, R.M.; Chakraverty, S. Residual Power Series Method for Solving Time-fractional Model of Vibration Equation of Large Membranes. *J. Appl. Comput. Mech.* **2019**, *5*, 603–615.
10. Jena, S.K.; Chakraverty, S. Differential Quadrature and Differential Transformation Methods in Buckling Analysis of Nanobeams. *Curved Layer. Struct.* **2019**, *6*, 68–76. [CrossRef]
11. Jena, S.K.; Chakraverty, S.; Jena, R.M.; Tornabene, F. A novel fractional nonlocal model and its application in buckling analysis of Euler-Bernoulli nanobeam. *Mater. Res. Express* **2019**, *6*, 1–17. [CrossRef]
12. Jena, S.K.; Chakraverty, S.; Tornabene, F. Vibration characteristics of nanobeam with exponentially varying flexural rigidity resting on linearly varying elastic foundation using differential quadrature method. *Mater. Res. Express* **2019**, *6*, 1–13. [CrossRef]
13. Jena, S.K.; Chakraverty, S.; Tornabene, F. Dynamical Behavior of Nanobeam Embedded in Constant, Linear, Parabolic and Sinusoidal Types of Winkler Elastic Foundation Using First-Order Nonlocal Strain Gradient Model. *Mater. Res. Express* **2019**, *6*, 0850f2. [CrossRef]
14. Jena, R.M.; Chakraverty, S.; Jena, S.K. Dynamic Response Analysis of Fractionally Damped Beams Subjected to External Loads using Homotopy Analysis Method. *J. Appl. Comput. Mech.* **2019**, *5*, 355–366.
15. Wang, C.M.; Zhang, Y.Y.; Ramesh, S.S.; Kitipornchai, S. Buckling analysis of micro-and nano-rods/tubes based on nonlocal Timoshenko beam theory. *J. Phys. D Appl. Phys.* **2006**, *39*, 3904. [CrossRef]
16. Emam, S.A. A general nonlocal nonlinear model for buckling of nanobeams. *Appl. Math. Model.* **2013**, *37*, 6929–6939. [CrossRef]
17. Yu, Y.J.; Xue, Z.N.; Li, C.L.; Tian, X.G. Buckling of nanobeams under nonuniform temperature based on nonlocal thermoelasticity. *Compos. Struct.* **2016**, *146*, 108–113. [CrossRef]
18. Nejad, M.Z.; Hadi, A.; Rastgoo, A. Buckling analysis of arbitrary two-directional functionally graded Euler–Bernoulli nano-beams based on nonlocal elasticity theory. *Int. J. Eng. Sci.* **2016**, *103*, 1–10. [CrossRef]
19. Dai, H.L.; Ceballes, S.; Abdelkefi, A.; Hong, Y.Z.; Wang, L. Exact modes for post-buckling characteristics of nonlocal nanobeams in a longitudinal magnetic field. *Appl. Math. Model.* **2018**, *55*, 758–775. [CrossRef]
20. Khaniki, H.B.; Hosseini-Hashemi, S. Buckling analysis of tapered nanobeams using nonlocal strain gradient theory and a generalized differential quadrature method. *Mater. Res. Express* **2017**, *4*, 065003. [CrossRef]
21. Yu, Y.J.; Zhang, K.; Deng, Z.C. Buckling analyses of three characteristic-lengths featured size-dependent gradient-beam with variational consistent higher order boundary conditions. *Appl. Math. Model.* **2019**, *74*, 1–20. [CrossRef]
22. Malikan, M. On the buckling response of axially pressurized nanotubes based on a novel nonlocal beam theory. *J. Appl. Comput. Mech.* **2019**, *5*, 103–112.
23. Malikan, M.; Dastjerdi, S. Analytical buckling of FG nanobeams on the basis of a new one variable first-order shear deformation beam theory. *Int. J. Eng. Appl. Sci.* **2018**, *10*, 21–34. [CrossRef]
24. Malikan, M.; Dimitri, R.; Tornabene, F. Transient response of oscillated carbon nanotubes with an internal and external damping. *Compos. Part B Eng.* **2019**, *158*, 198–205. [CrossRef]
25. Malikan, M.; Nguyen, V.B.; Tornabene, F. Damped forced vibration analysis of single-walled carbon nanotubes resting on viscoelastic foundation in thermal environment using nonlocal strain gradient theory. *Eng. Sci. Technol. Int. J.* **2018**, *21*, 778–786. [CrossRef]
26. Malikan, M.; Nguyen, V.B.; Dimitri, R.; Tornabene, F. Dynamic modeling of non-cylindrical curved viscoelastic single-walled carbon nanotubes based on the second gradient theory. *Mater. Res. Express* **2019**, *6*, 075041. [CrossRef]

27. Fazzolari, F.A. Generalized exponential, polynomial and trigonometric theories for vibration and stability analysis of porous FG sandwich beams resting on elastic foundations. *Compos. Part B Eng.* **2018**, *136*, 254–271. [CrossRef]

28. Fazzolari, F.A. Quasi-3D beam models for the computation of eigenfrequencies of functionally graded beams with arbitrary boundary conditions. *Compos. Struct.* **2016**, *154*, 239–255. [CrossRef]

29. Dimitri, R.; Tornabene, F.; Zavarise, G. Analytical and numerical modeling of the mixed-mode delamination process for composite moment-loaded double cantilever beams. *Compos. Struct.* **2018**, *187*, 535–553. [CrossRef]

30. Tornabene, F.; Fantuzzi, N.; Bacciocchi, M. Refined shear deformation theories for laminated composite arches and beams with variable thickness: Natural frequency analysis. *Eng. Anal. Bound. Elem.* **2019**, *100*, 24–47. [CrossRef]

31. Zakaria, M.; Al Harthy, A.M. Free Vibration of Pre-Tensioned Electromagnetic Nanobeams. *IOSR J. Math.* **2017**, *13*, 47–55. [CrossRef]

32. Reddy, J.N. Nonlocal theories for bending, buckling and vibration of beams. *Int. J. Eng. Sci.* **2007**, *45*, 288–307. [CrossRef]

33. Chakraverty, S. *Vibration of Plates*; CRC Press: Boca Raton, FL, USA, 2008.

34. Chakraverty, S.; Behera, L. *Static and Dynamic Problems of Nanobeams and Nanoplates*; World Scientific: Singapore, 2016.

 *nanomaterials*

Article

Modeling Analysis of Silk Fibroin/Poly(ε-caprolactone) Nanofibrous Membrane under Uniaxial Tension

Yunlei Yin [1,2], Xinfei Zhao [1] and Jie Xiong [1,3,*]

1   College of Materials and Textile, Zhejiang Sci-Tech University, Hangzhou 310018, China
2   School of Textile, Zhongyuan University of Technology, Zhengzhou 450007, China
3   Key Laboratory of Advanced Textile Materials and Manufacturing Technology of Ministry of Education, Zhejiang Sci-Tech University, Hangzhou 310018, China
*   Correspondence: jxiong@zstu.edu.cn; Tel.: +86-571-8684-3070; Fax: +86-571-8684-3082

Received: 4 July 2019; Accepted: 7 August 2019; Published: 10 August 2019

Abstract: Evaluating the mechanical ability of nanofibrous membranes during processing and end uses in tissue engineering is important. We propose a geometric model to predict the uniaxial behavior of randomly oriented nanofibrous membrane based on the structural characteristics and tensile properties of single nanofibers. Five types of silk fibroin (SF)/poly(ε-caprolactone) (PCL) nanofibers were prepared with different mixture ratios via an electrospinning process. Stress–strain responses of single nanofibers and nanofibrous membranes were tested. We confirmed that PCL improves the flexibility and ductility of SF/PCL composite membranes. The applicability of the analytical model was verified by comparison between modeling prediction and experimental data. Experimental stress was a little lower than the modeling results because the membranes were not ideally uniform, the nanofibers were not ideally straight, and some nanofibers in the membranes were not effectively loaded.

Keywords: electrospinning; nanofibrous membrane; geometric modeling; uniaxial tensile

---

1. Introduction

Electrospinning is an inexpensive and simple method that is broadly applicable for the controllable production of ultrafine continuous fibers with a high surface-to-volume ratio and high porosity. The tuning and controlling of these properties are often crucial for advanced biomedical applications like wound dressings [1,2], scaffold engineering [3,4], drug delivery devices [5], medical implants [6], and more [7,8].

Among the numerous materials suitable for tissue engineering, silk fibroin (SF) obtained from Bombyx mori silkworms is one of the most widely implemented as a scaffold material for tissue engineering due to its excellent biocompatibility and low immune reaction [9,10]. However, SF fibrous membranes fabricated via electrospinning are brittle due to the formation of a crystalline β-sheet secondary structure. This drawback limits the application of electrospun SF in tissue engineering [11]. Methods of improving SF mechanical performance include blending with other synthetic polymers such as polyethylene oxide (PEO) [12], poly(ε-caprolactone) (PCL) [13], polylactic acid (PLA) [14], polyglycolic acid (PGA) [15], and their copolymers [16,17]. The resulting nanocomposites possess the characteristics of the initial constituents, including the excellent mechanical properties of the polymers and the biocompatibility of SF [18]. SF is often combined with a biodegradable PCL given its outstanding strength and elasticity [19]. One example is preparation of electrospun SF/PCL nanofibrous scaffolds using formic acid (FA) [20] and 1,1,1,3,3,3-hexafluoro-2-propanol (HFIP) [19,21]. Applications of SF/PLC composites include regeneration skin [11], heart [22], bone [23] and vascular tissues [24].

Understanding the mechanics of nanofibrous membranes is important when evaluating their mechanical properties at various structural levels both during processing and use for final applications. Many researchers have tested the mechanical properties of polymer nanofibers and membranes. Ko et al. [25] obtained elastic moduli of carbon nanotube (CNT)/polyacrylonitrile (PAN) nanofibers using the atomic force microscopy (AFM) bending technique. Tan et al. [26] collected tensile strength data for a single electrospun PEO nanofiber using a mobile optical microscope stage coupled with a piezoresistive AFM tip. In another study, the tensile properties of a single-strand PCL electrospun ultrafine fiber were tested using a nano tensile instrument [27]. Lin et al. [28] reported a method based on a stream of air to determine the mechanical properties of electrospun fibers. Other researchers tested tensile properties of nanofibrous membranes [29–33]. The classical methods developed by Petterson [34] and Hearle and Stevenson [35] based on the mechanics of a nonwoven mat are still beneficial for understanding the mechanical properties of nanofibrous membranes. Yin et al. [36] analyzed how the properties of single nanofibers affect corresponding nanofibrous membrane.

Several analytical [37,38], semi-analytical [39,40] and numerical [41,42] methods have been adopted to predict the tensile properties of electrostatic textiles. However, these methods require a relatively complicated calculation. As the main factors affecting the mechanical properties of fibrous membranes are the micro-structure and single fiber properties of fibrous membranes, in this study we established the uniaxial tensile force relationship between single fibers and fibrous membranes using micro-mechanical analysis. Tensile mechanical properties of single fibers were correlated with those of fibrous membranes through a simple mechanical relationship model and the accuracy of the model was analyzed.

In this paper, a geometric modeling analysis is proposed to predict the uniaxial behaviors of randomly oriented nanofibrous membranes. For this purpose, we used the tensile and structural characteristics of single fibers and nanofibrous membrane, respectively. Five types of SF/PCL nanofibers with different mixture ratios were prepared via an electrospinning process. The stress–strain responses of single nanofibers from these nanofibrous membranes were tested for further use in the model. The applicability of the analytical model was examined by a comparison between the modeling prediction result and the experimental data.

2. Modeling Analysis

2.1. Assumptions

Firstly, the nanofibers in electrospun membrane were simplified as continuous and straight filaments. The membrane consisted of layers of randomly oriented nanofibers in the in-plane direction, as shown in Figure 1. Secondly, we excluded the interlayer effect. Nanofibers were deposited and randomly overlapped, ideally in sequence, during the whole electrospinning progress. No in-plane adhesion among fibers was considered. Therefore, the mechanical response of every nanofiber was assumed to be independent. Thirdly, time-dependent properties were not considered. The stretching speed was controlled to ensure that the specimens were under quasi-static tensile ($0.002\ \mathrm{s}^{-1}$). Neither the strain rate effect nor the stress relaxation were considered.

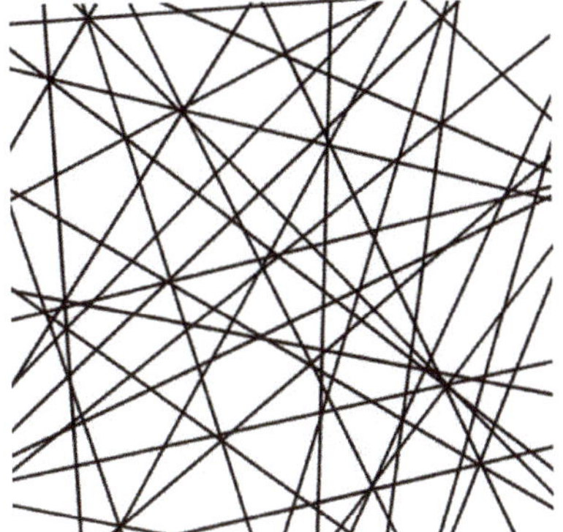

Figure 1. Schematic diagram of electrospun nanofiber membrane.

2.2. Derivation

The membrane consists of layers of randomly oriented nanofibers. Many intersections exist where fibers meet. As shown in Figure 2, a circle is used to represent a unit intersection of the membrane, and some randomly-oriented fibers pass through the center of the circle with the same length 2 r_0. Every fiber has its own angle (θ) of inclination. If a stretch along the y axis is applied, the circle becomes an ellipse after tensile stretching. Then, the angle of inclination (θ) becomes θ' and the length of fiber (2 r_0) becomes 2 r'. The stretch of the inclined fiber is coordinated with the deformation of the whole circle.

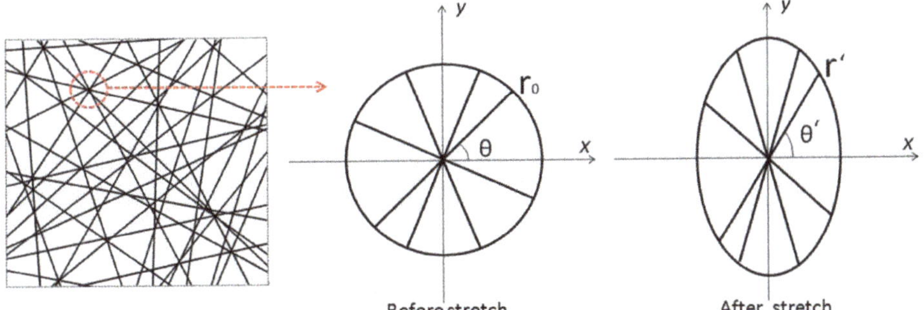

Figure 2. Changing of fiber's orientation under stretch.

Therefore, the strain of the membrane is described as follows:

$$\varepsilon = \frac{r'\sin\theta' - r_0\sin\theta}{r'\sin\theta'} \tag{1}$$

Due to the large deformation during stretch progress, the instant θ' is obtained by

$$\theta' = \arctan\frac{(1+\varepsilon)\sin\theta}{v\cos\theta} \tag{2}$$

where v is the shrinkage coefficients of fiber membranes, $v = \frac{\varepsilon_x}{\varepsilon_y}$. The deformed fibers are preferentially distributed along the stretching direction and fibrous membranes shrink horizontally. During tensile deformation, the shrinkage coefficient v of fibrous membranes is critical to the analysis of the deformation of nanofibers. Based on Petterson [43], since fibers are rearranged when stretching and the probability density function of the orientation distribution of fibers is difficult to calculate, the deformation calculation of single fibers is complicated. Therefore, it is feasible to determine the deformation of fibers by measuring and calculating the longitudinal and transverse deformations of fibers during stretching.

The force in the single fiber along the y axis by the applied loading is described as follows:

$$f_y = \frac{\pi}{4}d^2\sigma_f sin\theta' \, f_y = \frac{\pi}{4}d^2\sigma_f sin\theta' \tag{3}$$

where d is the diameter of the fiber and σ_f is the axial stress along the fiber's orientation. However, since the diameters of nanofibers are not uniform, the stress–strain data obtained from the single nanofiber tests should not be used directly to predict the mechanical behaviors of membranes. Therefore, the mathematical fitting, as shown in Figure 3 and Table 1, was applied to obtain stable results from the test data.

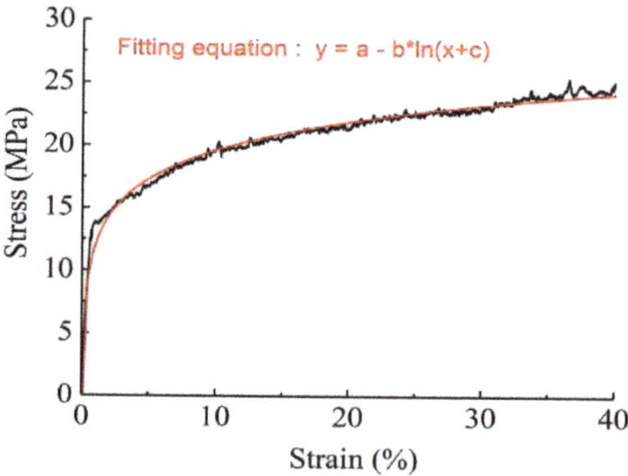

Figure 3. Typical stress–strain response of single nanofiber and its fitting equation.

Table 1. Parameters obtained from the fitting of single nanofibers.

Type	a [a]	b [a]	c [a]	R^2 [b]
SF/PCL = 100/0	27.207	−3.297	1.006×10^{-4}	0.984
SF/PCL = 75/25	22.669	−2.810	2.663×10^{-4}	0.976
SF/PCL = 50/50	26.576	−4.313	1.290×10^{-3}	0.975
SF/PCL = 25/75	12.569	−1.802	5.822×10^{-4}	0.949
SF/PCL = 0/100	17.130	−3.219	8.120×10^{-3}	0.975

[a] The fitting equation parameters for typical stress–strain response of single nanofiber were obtained from Supporting Information Figure 3. [b] Coefficient of determination.

Then the harmonic stress $\overline{\sigma_f}$ is based on the variation in the diameters of the nanofibers in membrane as follows:

$$\overline{\sigma_f} = \frac{1}{2}(1 + \frac{d}{D})[a - b\,ln(\varepsilon + c)] \tag{4}$$

where a, b, and c are fitting parameters, and D is the harmonic average of the diameters of the nanofibers. The fibers in the membrane are randomly oriented. Therefore, the integral algorithm was used to obtain the average force $\overline{f_y}$ along the y axis from every fiber with oriented angle from 0 to $\pi/2$ as follows:

$$\overline{f_y} = \frac{2}{\pi}\int_0^{\frac{\pi}{2}} f_y d(\theta) \tag{5}$$

Finally, the total force F_y along the y axis from all fibers is described as

$$F_y = n\overline{f_y} \tag{6}$$

$$n = \frac{2V(1-p)}{r_0 \pi D^2} \qquad (7)$$

where n is the amount of fibers in unit volume V and p is the porosity of the electrospun membrane.

3. Materials and Methods

3.1. Materials

SF/PLC spinning dope solution was prepared using regenerated SF sponge and PCL (with molecular weight of ~80,000 g·mol^{-1}, Guanghua Weiye Co., Ltd., Shenzhen, China). Other materials (Na$_2$CO$_3$, CaCl$_2$ and 99% methanol) were obtained from Hangzhou Gaojing Fine Chemicals Co., Ltd. (Hangzhou, China). Formic acid (FA. 98.0% pure) was purchased from Shanghai Lingfeng Chemical Reagent Co., Ltd. (Shanghai, China). All materials were used as received without any additional treatment.

3.2. Preparation of Stock SF Sponge

Silk cocoons were boiled in 0.05 vol% Na$_2$CO$_3$ solution for 30 min, after which they were rinsed with distilled water and dried at 40 °C overnight. After that, these degummed silk threads were placed into a 1:8:2 mixture (by moles of CaCl$_2$, H$_2$O, and EtOH) at 75 °C for 5 min. After the silk threads dissolved, they were dialyzed using a cellulose 12–14 kDa tubing. Distilled water was used as counter-solution. Dialysis lasted for 3 days. The resulting aqueous SF solution was removed from the dialysis tubing, filtered and freeze dried. The resulting materials were regenerated SF sponges, which were stored in a desiccator prior to their use.

3.3. Preparation of SF/PCL Solutions

Mixtures containing regenerated SF sponges and PCL at 100/0, 75/25, 50/50, 25/75, and 0/100 weight ratios were dissolved in 18 wt % FA solution and maintained at room temperature for 2 h. After that, the mixtures were stirred for 3 h.

3.4. Preparation of SF/PCL Nanofibrous Membranes

A syringe with a stainless-steel needle (22 G) containing 10 mL of spinning dope solution was placed into a special pump. Electrospinning conditions were 0.6 mL·h^{-1} feed rate and 15 kV voltage applied between the needle tip and an aluminum sheet collector located 10 cm away from the needle and mounted on the vertical metal mesh surface. Relative humidity was maintained below 50% during electrospinning.

3.5. Morphology of the Prepared Nanofibrous Membranes

Nanofiber morphologies were characterized by a Carl Zeiss (Oberkochen, Germany) field emission scanning electron microscopy (FE-SEM) instrument operated at 3 kV accelerated voltage. A thin gold layer was sputtered on the fibers to improve their conductivity. The average diameter was calculated based on 100 measurements of different fibers, which were recorded using Image-Pro Plus 6.2 software (ICube, Crofton, MD, USA) using FE-SEM images. Micrographs and diameters of nanofibers from five different kinds of membranes are presented in Figure 4 and Table 2, respectively.

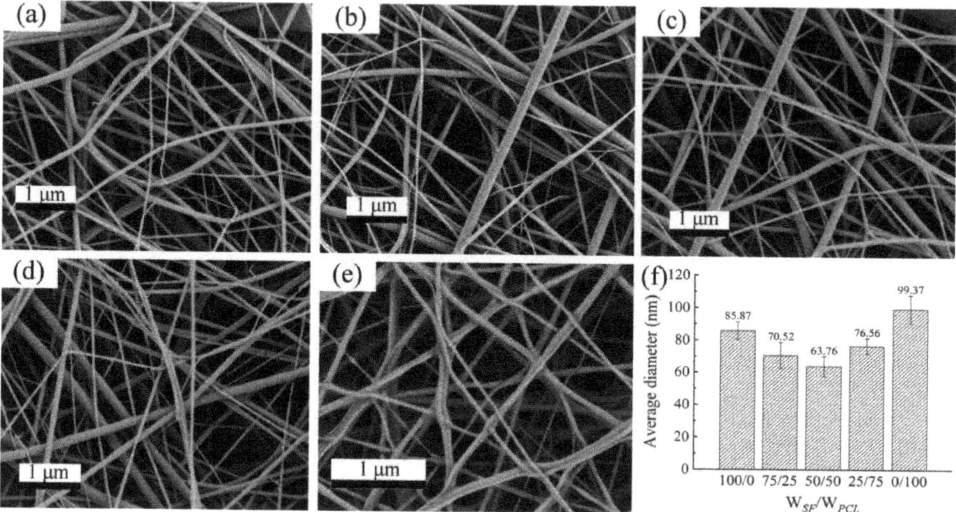

Figure 4. Field emission (FE)-SEM micrographs of the five different types of electrospun membranes: (**a**) $W_{SF}{:}W_{PCL} = 100{:}0$, (**b**) $W_{SF}{:}W_{PCL} = 75{:}25$, (**c**) $W_{SF}{:}W_{PCL} = 50{:}50$, (**d**) $W_{SF}{:}W_{PCL} = 25{:}75$, and (**e**) $W_{SF}{:}W_{PCL} = 0{:}100$; and (**f**) average diameter of nanofibers.

Table 2. Nanofiber diameters in membranes.

Type	Arithmetic Average Diameter (nm)	Root-Mean-Square Diameter (nm)	Harmonic Average of Diameter (nm)
SF/PCL = 100/0	85.87	91.29	79.49
SF/PCL = 75/25	70.52	76.12	62.96
SF/PCL = 50/50	63.76	71.91	57.11
SF/PCL = 25/75	76.56	80.41	70.54
SF/PCL = 0/100	99.37	110.18	88.62

3.6. Tensile Test for Single Nanofiber and Nanofibrous Membranes

Stress along the axial of a single nanofiber was tested using a nano-mechanical stretching system (Agilent UTM T150, Santa Clara, CA, USA), which provides excellent mechanical characterization at the nano-scale due to its unique actuating transducer that is capable of generating tensile force load on individual fibers by electromagnetic actuation combined with a precise capacitive gauge (Figure 5).

The uniaxial stress–strain curve of the nanofibrous membranes were tested using the KES-G1 tensile system (Kato-Tech Company, Kyoto, Japan) as shown in Figure 6. Specimens were fixed by two overlap-pasted frames composed of paper to avoid incline and torsion. Before testing, the left and right arms of the paper frames were cut to avoid affecting the stretching of the specimen during tensile testing.

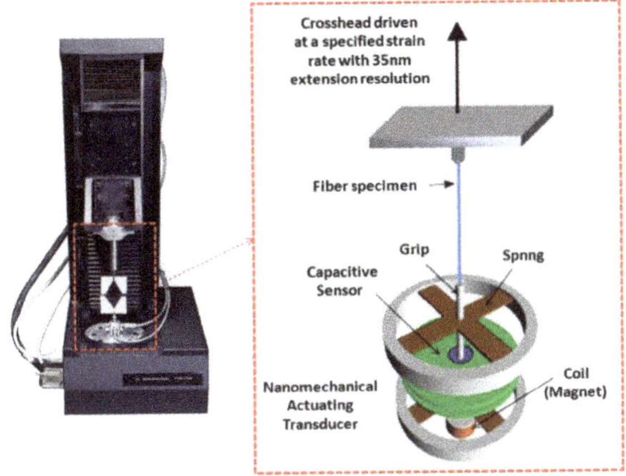

Figure 5. Tensile testing system for single nanofibers.

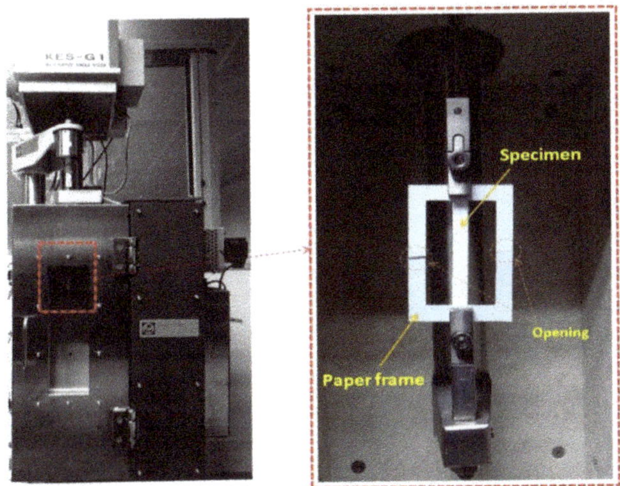

Figure 6. Tensile testing system for nanofibrous membranes.

4. Results and Discussion

Figure 7 demonstrates the typical deformative characteristics of nanofibrous membranes under uniaxial tensile testing. Membranes were stretched until fracture occurred. We observed that the rectangle specimen was forced into a dog-bone shape due to tension in the membrane before fracture. The specimen catastrophically failed with a triangular cracking area from the edge of the membrane, which indicated the ductile deformation of nanofibers in the membrane.

Figure 8 shows the uniaxial stress–strain curves of the five different types of SF/PCL membranes. Post-fracture behaviors were not retained for all curves. We found that the strain of membranes increased with increasing PCL content in nanofibers. This indicates that PCL helped to improve the flexibility and ductility of the composite membranes, and that the membranes' strengths were not obviously affected. These properties are critical for materials used for engineering of vascular and skin tissues.

By comparing the modeling prediction and experimental data, we found that the analytical model is somewhat applicable. The proposed modeling prediction demonstrated similarity to the testing results, showing that the mechanical behaviors of membranes could be predicted based on their structural characteristics and the properties of single nanofibers. However, the experimentally recorded stress values were a little lower than the modeling values. The morphology of the membranes could explain this difference. Firstly, the nanofibrous membranes were not ideally uniform, which could lead to the existence of some weak regions. Secondly, nanofibers were not ideally straight, which could lead to a slow initial increase in stress due to the adjustment of the nanofibers themselves. Thirdly, some nanofibers in membranes were potentially not loaded during tensile process.

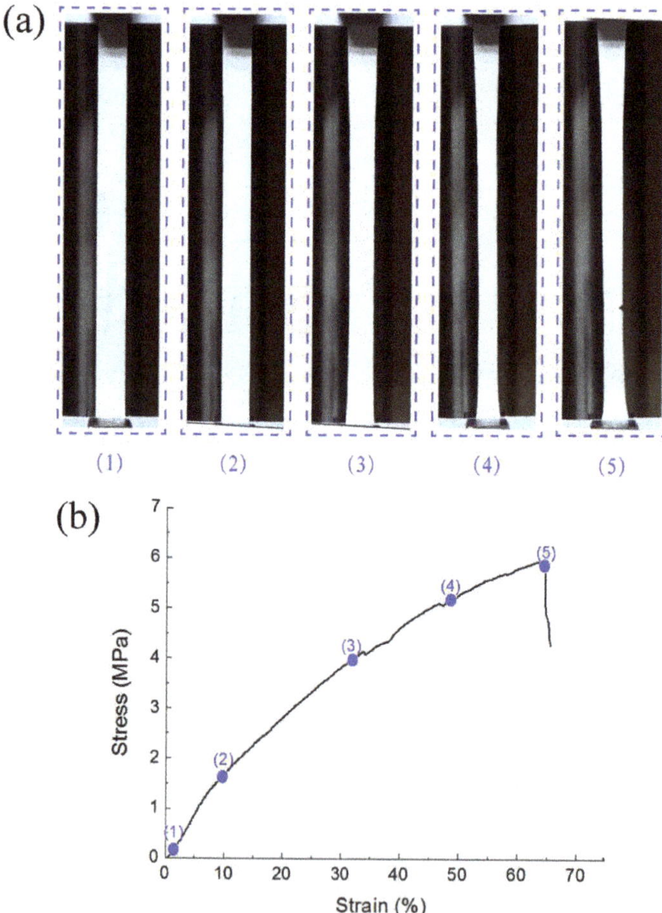

Figure 7. Typical deformation characteristics of nanofibrous membrane with the progress of stress–strain response under uniaxial tensile testing: (**a**) morphology changes of nanofibrous membranes, (**b**) stress–strain curves of nanofibrous membranes.

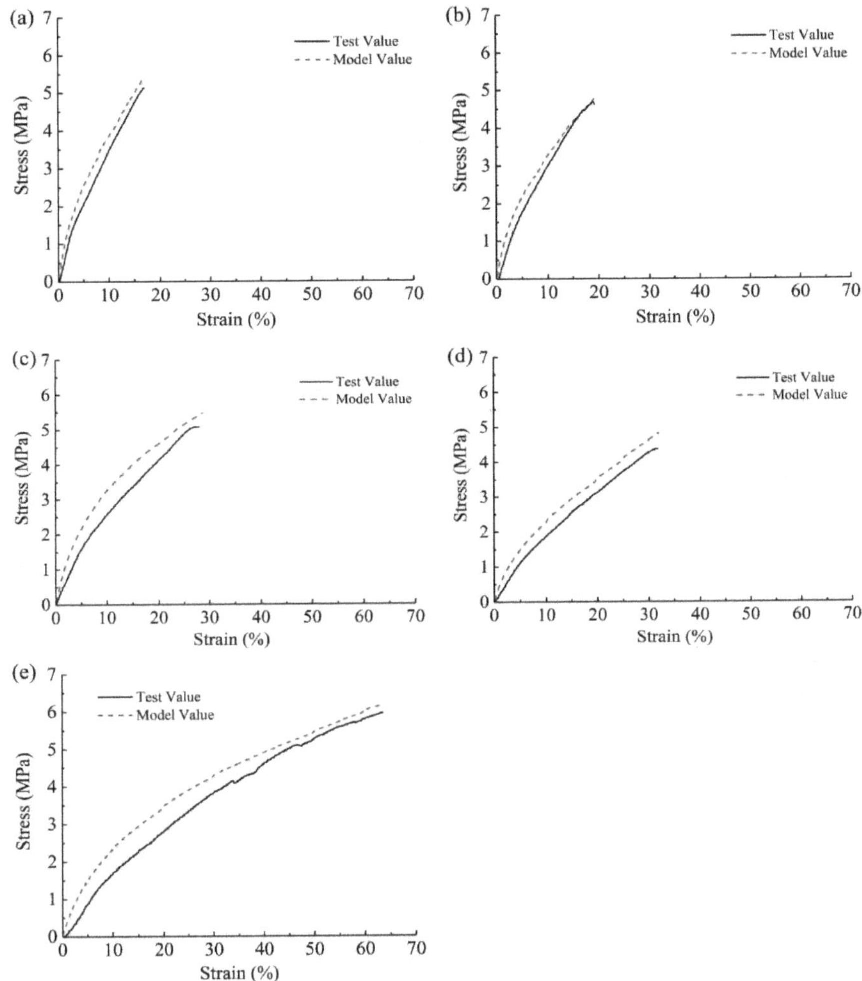

Figure 8. Results of the modeling analysis and experimental tests: (**a**) W_{SF}:W_{PCL} = 100:0, (**b**) W_{SF}:W_{PCL} = 75:25, (**c**) W_{SF}:W_{PCL} = 50:50, (**d**) W_{SF}:W_{PCL} = 25:75, and (**e**) W_{SF}:W_{PCL} = 0:100 membranes.

5. Conclusions

SF/PCL nanofibrous membranes were produced using an electrospinning technique using formic acid. The presence of PCL helped to improve the flexibility and ductility of the membrane without compromising the strength.

In this study, we proposed a geometric modeling analysis based on the tensile properties of single fibers and the structural characteristics of nanofibrous membranes. The applicability of the analytical model was verified by a comparison between the model prediction and experimental data. The experimentally recorded stress level was a little lower than the modeling results for three reasons: (1) the nanofibrous membranes were not ideally uniform, (2) the nanofibers were not ideally straight, and (3) some nanofibers in the membranes were not effectively loaded.

Author Contributions: J.X. obtained the research funding; Y.Y. conceived the study, made charts and drafted the manuscript; X.Z. designed the manuscript framework and performed the experiments.

Funding: This research was funded by the National Natural Science Foundation of China (Grant No. 11272289), the Zhejiang Provincial Natural Science Foundation of China (Grant No. LZ16E020002), the Program for Innovative Research Team of Zhejiang Sci-Tech University (No. 15010039-Y).

Acknowledgments: We greatly acknowledge the support by the Research Foundation for Young Doctor Teachers of Zhongyuan University of Technology.

Conflicts of Interest: The authors declare no conflict of interest.

References

1. Liu, M.; Duan, X.P.; Li, Y.M.; Yang, D.P.; Long, Y.Z. Electrospun nanofibers for wound healing. *Mater. Sci. Eng. C* **2017**, *76*, 1413–1423. [CrossRef] [PubMed]

2. Nosratollah, Z.; Roghayeh, S.; Amir, F.; Abbas, M.; Mehdi, D.; Younes, P.-S. An Overview on Application of Natural Substances Incorporated with Electrospun Nanofibrous Scaffolds to Development of Innovative Wound Dressings. *Mini-Rev. Med. Chem.* **2017**, *17*, 1–14.

3. Dong, C.; Lv, Y. Application of collagen scaffold in tissue engineering: Recent advances and new perspectives. *Polymers* **2016**, *8*, 42. [CrossRef] [PubMed]

4. Salifu, A.A.; Lekakou, C.; Labeed, F.H. Electrospun oriented gelatin-hydroxyapatite fiber scaffolds for bone tissue engineering. *J. Biomed. Mater. Res. Part A* **2017**, *105*, 1911–1926. [CrossRef] [PubMed]

5. Weng, L.; Xie, J. Smart electrospun nanofibers for controlled drug release: Recent advances and new perspectives. *Curr. Pharm. Des.* **2015**, *21*, 1944–1959. [CrossRef] [PubMed]

6. Gnavi, S.; Fornasari, B.E.; Tonda-Turo, C.; Laurano, R.; Zanetti, M.; Ciardelli, G.; Geuna, S. The Effect of Electrospun Gelatin Fibers Alignment on Schwann Cell and Axon Behavior and Organization in the Perspective of Artificial Nerve Design. *Int. J. Mol. Sci.* **2015**, *16*, 12925–12942. [CrossRef] [PubMed]

7. Zou, B.; Guo, Y.; Shen, N.; Xiao, A.; Li, M.; Zhu, L.; Wan, P.; Sun, X. Sulfophenyl-Functionalized Reduced Graphene Oxide Networks on Electrospun 3D Scaffold for Ultrasensitive NO_2 Gas Sensor. *Sensors* **2017**, *17*, 2954. [CrossRef]

8. Huan, S.; Liu, G.; Cheng, W.; Han, G.; Bai, L. Electrospun Poly(lactic acid)-Based Fibrous Nanocomposite Reinforced by Cellulose Nanocrystals: Impact of Fiber Uniaxial Alignment on Microstructure and Mechanical Properties. *Biomacromolecules* **2018**, *19*, 1037–1046. [CrossRef]

9. Cho, H.J.; Ki, C.S.; Oh, H.; Lee, K.H.; Um, I.C. Molecular weight distribution and solution properties of silk fibroins with different dissolution conditions. *Int. J. Biol. Macromol.* **2012**, *51*, 336–341. [CrossRef]

10. Park, S.Y.; Ki, C.-S.; Park, Y.H.; Jung, H.M. Electrospun Silk Fibroin Scaffolds with Macropores for Bone Regeneration: An In Vitro and In Vivo Study. *Tissue Eng. Part A* **2010**, *16*, 1271–1279. [CrossRef]

11. Chutipakdeevong, J.; Ruktanonchai, U.; Supaphol, P. Hybrid biomimetic electrospun fibrous mats derived from poly(ε-caprolactone) and silk fibroin protein for wound dressing application. *J. Appl. Polym. Sci.* **2014**, *132*, 41653. [CrossRef]

12. Chen, W.; Li, D.; EI-Shanshory, A.; EI-Newehy, M.; EI-Hamshary, H.A.; AI-Deyab, S.S.; He, C.L.; Mo, X. Dexamethasone loaded core-shell SF/PEO nanofibers via green electrospinning reduced endothelial cells inflammatory damage. *Colloids Surf. B Biointerfaces* **2015**, *126*, 561–568. [CrossRef] [PubMed]

13. Li, W.; Qiao, X.; Sun, K.; Chen, X. Mechanical and viscoelastic properties of novel silk fibroin fiber/poly(ε-caprolactone) biocomposites. *J. Appl. Polym. Sci.* **2008**, *110*, 134–139. [CrossRef]

14. Buasri, A.; Chaiyut, N.; Loryuenyong, V.; Jaritkaun, N. Mechanical and thermal properties of silk fiber reinforced poly(lactic acid) biocomposites. *Optoelectron. Adv. Mater.-Rapid Commun.* **2013**, *7*, 938–942.

15. Zhang, J.; Mo, X. Current research on electrospinning of silk fibroin and its blends with natural and synthetic biodegradable polymers. *Front. Mater. Sci.* **2013**, *7*, 129–142. [CrossRef]

16. Zhang, K.; Qing, Y.; Yan, Z. Degradation of electrospun SF/P (LLA-CL) blended nanofibrous scaffolds in vitro. *Polym. Degrad. Stab.* **2011**, *96*, 2266–2275. [CrossRef]

17. Zhang, K.; Wang, H.; Huang, C.; Su, Y. Fabrication of silk fibroin blended P (LLA-CL) nanofibrous scaffolds for tissue engineering. *J. Biomed. Mater. Res. Part A* **2010**, *93*, 984–993. [CrossRef]

18. Gunn, J.; Zhang, M. Polyblend nanofibers for biomedical applications: Perspectives and challenges. *Trends Biotechnol.* **2010**, *28*, 189–197. [CrossRef]

19. Yin, Y.; Xiong, J. Effect of the Distribution of Fiber Orientation on the Mechanical Properties of Silk Fibroin/Polycaprolactone Nanofiber Mats. *J. Eng. Fibers Fabr.* **2016**, *12*, 17–28. [CrossRef]

20. Yuan, H.; Shi, H.; Qiu, X.; Chen, Y. Mechanical property and biological performance of electrospun silk fibroin-polycaprolactone scaffolds with aligned fibers. *J. Biomater. Sci. Polym. Ed.* **2016**, *27*, 1–24. [CrossRef]

21. Lv, J.; Chen, L.; Zhu, Y.; Hou, L.; Liu, Y. Promoting epithelium regeneration for esophageal tissue engineering through basement membrane reconstitution. *ACS Appl. Mater. Interfaces* **2014**, *6*, 4954–4964. [CrossRef]

22. Sridhar, S.; Venugopal, J.R.; Sridhar, R.; Ramakrishna, S. Cardiogenic differentiation of mesenchymal stem cells with gold nanoparticle loaded functionalized nanofibers. *Colloids Surf. B-Biointerfaces* **2015**, *134*, 346–354. [CrossRef]

23. Kim, B.S.; Park, K.E.; Kim, M.H.; You, H.K.; Lee, J.; Park, W.H. Effect of nanofiber content on bone regeneration of silk fibroin/poly(ε-caprolactone) nano/microfibrous composite scaffolds. *Int. J. Nanomed.* **2015**, *10*, 485–502.

24. Wolfe, P.S.; Madurantakam, P.; Garg, K.; Sell, S.A.; Beckman, M.J.; Bowlin, G.L. Evaluation of thrombogenic potential of electrospun bioresorbable vascular graft materials: Acute monocyte tissue factor expression. *J. Biomed. Mater. Res. Part A* **2010**, *92*, 1321–1328. [CrossRef]

25. Ko, F.; Gogotsi, Y.; Ali, A.A.; Naguib, N. Electrospinning of Continuous Carbon Nanotube-Filled Nanofiber Yarns. *Adv. Mater.* **2003**, *15*, 1161–1165. [CrossRef]

26. Tan, E.P.S.; Goh, C.N.; Sow, C.H.; Lim, C.T. Tensile test of a single nanofiber using an atomic force microscope tip. *Appl. Phys. Lett.* **2005**, *86*, 073115. [CrossRef]

27. Tan, E.P.S.; Ng, S.Y.; Lim, C.T. Tensile testing of a single ultrafine polymeric fiber. *Biomaterials* **2005**, *26*, 1453–1456. [CrossRef]

28. Lin, Y.; Clark, D.M.; Reneker, D.H. Mechanical properties of polymer nanofibers revealed by interaction with streams of air. *Polymer* **2012**, *53*, 782–790. [CrossRef]

29. Lee, K.H.; Kim, H.Y.; Khil, M.S.; Ra, Y.M.; Lee, D.R. Characterization of nano-structured poly(ε-caprolactone) nonwoven mats via electrospinning. *Polymer* **2003**, *44*, 1287–1294. [CrossRef]

30. Huang, Z.M.; Zhang, Y.Z.; Ramakrishna, S.; Lim, C.T. Electrospinning and mechanical characterization of gelatin nanofibers. *Polymer* **2004**, *45*, 5361–5368. [CrossRef]

31. Ayutsede, J.; Gandhi, M.; Sukigara, S.; Micklus, M.; Chen, H.E.; Ko, F. Regeneration of Bombyx mori silk by electrospinning. Part 3: Characterization of electrospun nonwoven mat. *Polymer* **2005**, *46*, 1625–1634. [CrossRef]

32. Qiang, J.; Wan, Y.; Yang, L.; Cao, Q. Effect of Ultrasonic Vibration on Structure and Performance of Electrospun PAN Fibrous Membrane. *J. Nano Res.* **2013**, *23*, 96–103. [CrossRef]

33. Cao, Q.; Wan, Y.; Qiang, J.; Yang, R.; Fu, J.; Wang, H.; Gao, W.; Ko, F. Effect of sonication treatment on electrospinnability of high-viscosity PAN solution and mechanical performance of microfiber mat. *Iran. Polym. J.* **2014**, *23*, 1–7. [CrossRef]

34. Petterson, D.R. Mechanics of Nonwoven Fabrics. *Ind. Eng. Chem.* **1959**, *51*, 902–903. [CrossRef]

35. Hearle, J.W.S.; Stevenson, P.J. Studies in Nonwoven Fabrics Part IV: Prediction of Tensile Properties1. *Text. Res. J.* **1964**, *34*, 181–191. [CrossRef]

36. Yin, Y.; Pu, D.; Xiong, J. Analysis of the Comprehensive Tensile Relationship in Electrospun Silk Fibroin/Polycaprolactone Nanofiber Membranes. *Membranes* **2017**, *7*, 67. [CrossRef]

37. Wan, L.Y.; Wang, H.; Gao, W.; Ko, F. An analysis of the tensile properties of nanofiber mats. *Polymer* **2015**, *73*, 62–67. [CrossRef]

38. Pai, C.L.; Boyce, M.C.; Rutledge, G.C. On the importance of fiber curvature to the elastic moduli of electrospun nonwoven fiber meshes. *Polymer* **2011**, *52*, 6126–6133. [CrossRef]

39. Rizvi, M.S.; Pal, A. Statistical model for the mechanical behavior of the tissue engineering non-woven fibrous matrices under large deformation. *J. Mech. Behav. Biomed. Mater.* **2014**, *37*, 235–250. [CrossRef]

40. Ban, E.; Barocas, V.H.; Shephard, M.S.; Picu, C.R. Effect of fiber crimp on the elasticity of random fiber networks with and without embedding matrices. *J. Appl. Mech.* **2016**, *83*, 0410081–0410087. [CrossRef]

41. Silberstein, M.N.; Pai, C.L.; Rutledge, G.C.; Boyce, M.C. Elastic-plastic behavior of non-woven fibrous mats. *J. Mech. Phys. Solids* **2012**, *60*, 295–318. [CrossRef]

42. Broedersz, C.P.; Mao, X.; Lubensky, T.C.; MacKintosh, F.C. Criticality and isostaticity in fibre networks. *Nat. Phys.* **2011**, *7*, 983–988. [CrossRef]
43. Petterson, D.R. *On the Mechanics of Non-Woven Fabrics*; Massachusetts Institute of Technology: Cambridge, MA, USA, 1959.

Article

Detwinning Mechanism for Nanotwinned Cubic Boron Nitride with Unprecedented Strength: A First-Principles Study

Bo Yang [1], Xianghe Peng [1,2,*], Sha Sun [1], Cheng Huang [1], Deqiang Yin [1], Xiang Chen [3] and Tao Fu [1,*]

1   Department of Engineering Mechanics, Chongqing University, Chongqing 400044, China
2   State Key Laboratory of Coal Mining Disaster Dynamics and Control, Chongqing University, Chongqing 400044, China
3   Advanced Manufacturing Engineering, Chongqing University of Posts and Telecommunications, Chongqing 400065, China
*   Correspondence: xhpeng@cqu.edu.cn (X.P.); futao@cqu.edu.cn (T.F.)

Received: 12 July 2019; Accepted: 26 July 2019; Published: 3 August 2019

Abstract: Synthesized nanotwinned cubic boron nitride (nt-cBN) and nanotwinned diamond (nt-diamond) exhibit extremely high hardness and excellent stability, in which nanotwinned structure plays a crucial role. Here we reveal by first-principles calculations a strengthening mechanism of detwinning, which is induced by partial slip on a glide-set plane. We found that continuous partial slip in the nanotwinned structure under large shear strain can effectively delay the structural graphitization and promote the phase transition from twin structure to cubic structure, which helps to increase the maximum strain range and peak stress. Moreover, ab initio molecular dynamics simulation reveals a stabilization mechanism for nanotwin. These results can help us to understand the unprecedented strength and stability arising from the twin boundaries.

Keywords: nanotwin; detwinning; extreme hardness; excellent stability

---

1. Introduction

Diamond and zinc blende structured materials are the important members in the family of superhard materials, among which diamond and cubic boron nitride (cBN) are the most prominent representatives. These strong covalent bond solids are indispensable to fundamental scientific research and technological applications in many fields [1–8]. With the development of synthesis technology at nanoscale, more and more attention has been paid to nanocrystalline (NC) superhard materials as well as their outstanding mechanical properties [9–13]. It has been found that grain size plays a decisive role in the strength of materials [14–16]. For example, the strength of NC Cu increases with the decrease of grain size, and reaches the maximum at the critical size ($d = 19.3$ nm), followed by softening with the further decrease of grain size [17]. Similar phenomena have also been observed in covalently bonded materials, for example, at room temperature the Knoop hardness of an NC diamond with grain size of 10–20 nm is about 130 GPa, much higher than that of single crystal diamond, which is about 70–90 GPa [18,19], and the strength of NC cBN increases with the decrease of grain size d [20,21], following the famous Hall-Petch relationship [15,22].

Recent studies [23] showed that nanotwinned cBN (nt-cBN) and nanotwinned diamond (nt-diamond) exhibit extremely high mechanical and thermal properties compared with their single crystal counterparts. The Vickers hardness of nt-cBN with an average twin thickness of $\lambda = 3.8$ nm reaches 95–108 GPa, which even exceeds that of a single crystal diamond (90 GPa) [23]. This high hardness was attributed to the existence of high-density nanotwins [24]. On the other hand, the onset oxidation temperature of nt-cBN is also about 200 °C higher than that of single cBN crystal. The corresponding studies [23,25] challenge the general understanding of the properties of materials at nanoscale. It has been shown that, in nanotwinned metals [26–28], the appropriate thickness and distribution of twins play significant roles in the improvements of the physical and mechanical properties of metals, including strength, hardness, and thermal stability. However, it has been shown in many studies that the strengthening/weakening mechanisms between nt-metals and nt-cBN are obviously different [24,29,30]. At present, the role of nanotwins in the enhancement of the mechanical properties of strong covalent bond solids remains controversial [31,32] and needs to be clarified.

In this work, the mechanisms for the enhancement of mechanical properties and stability of nt-cBN under continuous shear strain are investigated using first-principles calculations. The paper is organized as follows: the calculation details are briefly introduced in the following section, followed by the presentation of the results calculated and the corresponding discussions, and some conclusions are drawn and shown in the last section.

2. Methods

In this work, the stability and mechanical properties of materials are described by the stress-strain relationships, twin boundary energies (TBEs) and ab initio molecular dynamics (MD) simulations, respectively. All simulations are performed using the Vienna ab initio simulation package (VASP) code based on the density functional theory (DFT) with the generalized gradient approximation of Perdew-Burke-Ernzerh of version [33] for the exchange-correlation energy and a plane-wave basis set [34]. The projector augmented wave (PAW) method describes the electron-ion interaction [35], in which $2s^22p^1$, $2s^22p^2$, and $2s^22p^3$ are the valence electrons for B, C, and N atoms, respectively. An energy cutoff of 500 eV and Monkhorst-Pack k-point grids [36] of $5 \times 7 \times 3$ are used for the calculation of the differences in the total energy and stress response of perfect crystal and twin structure. The convergence criterion is that the total energy of the system and the force on each of the atoms are less than 1×10^{-4} eV and 0.001 eV/Å, respectively.

We utilize standard ab initio MD simulations as implemented in the VASP code to observe insights into the equilibrium structure of diamond and cBN at a given temperature. The setting accuracy of calculation parameters is consistent with that of the basic properties of the above DFT energy calculation. Moreover, in the intermediate period, a micro-canonical ensemble (NVE) is simulated. A supercell containing 96 atoms is employed to avoid artifacts associated with constraints imposed by finite-sized unit cells [37]. In addition, the parameters related to the simulation time are set to 3 femtoseconds (fs) per time step and the maximum ionic step is 1500.

To obtain the failure shear stress, we apply a shear strain component on a crystal cell along the prescribed orientation, while relaxing the other five strain components, and deform the periodic model until it fails [7,38–40]. To determine the lowest energy path during deformation, we calculate the generalized stacking fault energy (GSFE) surface, also called γ-surface, which describes the energy variation between that of the perfect crystal and that of the half-crystal shift d on the crystallographic plane prescribed [41]. For example, to obtain the γ-surface of the (111) plane of cBN, we divide the plane of cBN into 525 grids along the $[11\bar{2}]$ and $[1\bar{1}0]$ directions, calculate the difference between the energy of the perfect crystal and that with the origin of the upper-half shifted by the distance to each grid.

3. Results and Discussions

3.1. Excellent Stability of Nanotwinned Structure

Figure 1 shows schematically the atomic arrangement of perfect crystal and twin of cBN. In order to measure the stability of the twin structure, we calculate the twin boundary energy (TBE), which represents the energy variation between the perfect crystal and twin structure of per unit area.

The smaller the TBE, the better the stability of the twin structure [38]. The TBEs of cBN and diamond are calculated and listed in Table 1, where one can see that the TBEs of cBN and diamond are 81.72 and 101.69 mJ/m^2, respectively. Moreover, the total energy of crystalline cBN and that of twined cBN are basically the same, so are those of diamond. The above information indicates that twin structure should be as stable as its single crystal for cBN and diamond. However, the extremely high GSFE may limit the formation of the twin structure in cBN. The recent report [23,25] of nt-cBN converted from cBN nano-onions with high-density defects under high temperatures and pressures solves this synthetic problem.

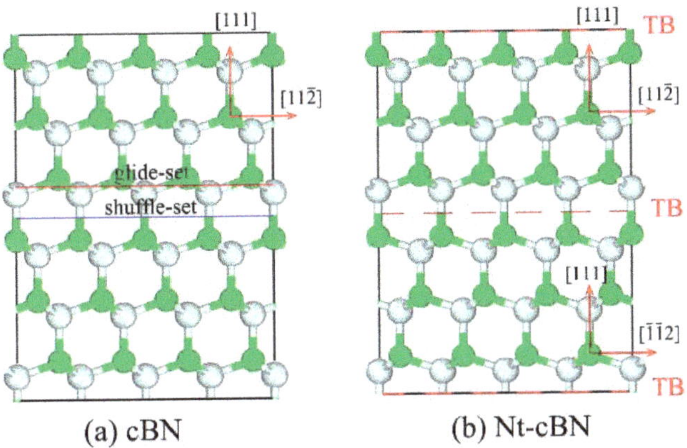

Figure 1. Atomic arrangement of cubic boron nitride (cBN) and twin cBN: (**a**) cBN, with two non-equivalent (111) planes, corresponding respectively to narrowly spaced atomic layer (glide-set) indicated by red line, and widely spaced atomic layer (shuffle-set) by blue line [42]. (**b**) nanotwinned cubic boron nitride (nt-cBN) with red lines as twin boundary (TBs).

Table 1. Calculated total energies and TBEs of cBN and diamond.

	Total Energy (ev)		TBE (mJ/m^2)
	Perfect Crystal	**Twin**	
cBN	−836.29	−835.83	81.72
Diamond	−872.62	−872.06	101.69

To reveal the contribution of the nanotwinned structure to the unusual thermal stability of the strong covalent bond solid, ab initio MD simulations were carried out. We have qualitatively compared the thermal stability of diamond and nt-diamond at T = 1500 K. Results of our ab initio MD simulations are presented as structural snapshots of diamond and nt-diamond at T = 1500 K in Figure 2a,b. In addition, the Supplementary Movies S1 and S2 respectively describe ab initio molecular dynamics simulations of diamond and nt-diamond at T = 1500 K in the front view. We adopted a covalent bond cutoff radius of 1.5 Å for diamond and nt-diamond. As shown in Figure 2a,b, some C–C bonds of the diamond are broken while the C–C bonds of nt-diamond are basically intact. It is obvious that the structural change of nt-diamond is less than that of the diamond, indicating the twin structure does contribute to the stability of the diamond, which verifies the conclusion of the experiment. It should be noted that structural change refers to the difference between a new structure and the original structure. In parallel, the twin structure does not contribute significantly to the thermal stability of cBN, which may be related to the superior thermal stability of cBN itself [43,44]. Figure 2c,d, and Supplementary Movies S3 and S4 give more information.

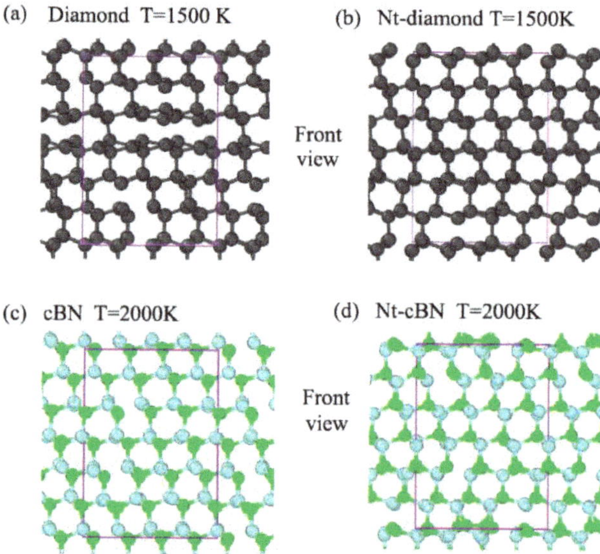

Figure 2. (**a**) and (**b**) are snapshots of ab initio molecular dynamics simulations depicting structural changes in diamond and nt-diamond at T = 1500 K. Corresponding snapshots of cBN and nt-cBN at T = 2000 K are shown in (**c**) and (**d**). The solid purple line represents unit cell.

It has been reported in the previous studies [45,46] that the (111) cleavage plane clearly dominates the fracture of cBN or diamond. The γ-surfaces in cBN for glide-set and shuffle-set planes are calculated for the determination of the lowest energy path during deformation, as shown in Figure 3a,b, respectively. It can be seen that the lowest energy path of $\frac{a_0}{6}[11\bar{2}]$ ($a_0 = 3.625$ Å, representing the lattice constant) for glide-set and shuffle-set planes, where the peak of the path is defined as the energy barrier (γ_U). Then, the GSFE curves for glide-set and shuffle-set planes in (111)<112> slip system are extracted and shown in Figure 4a, where the γ_{Ug} on the glide-set plane is 3.72 J/m², which is smaller than that on the shuffle-set plane (6.85 J/m²), indicating that the slip of cBN should be dominated by the partial slip on the glide-set plane, which coincides with the conclusion that the stacking fault has been observed experimentally [47], as shown in Figure 3c–f. In order to understand more clearly the stacking faults in the experiment, the atomic configurations of several key structural points have been revealed. As shown in Figure 4b, the stacking sequence of the (111) planes of cBN can be expressed as $A\alpha B\beta C\gamma A\alpha B\beta C\gamma \dots$. The upper part of the model is rigidly displaced along the $[11\bar{2}]$ direction on the glide-set plane. Figure 4c shows the atomic configuration when normalized displacement is equal to $\frac{a_0}{12}[11\bar{2}]$, which is unstable, corresponding to the unstable GSFE of γ_{Ug}. When normalized displacement is equal to $\frac{a_0}{6}[11\bar{2}]$, as shown in Figure 4d, the stacking sequences becomes $A\alpha B\beta C\alpha B\beta C\gamma A\alpha \dots$, indicating the occurrence of a stacking fault. At this moment, the GSFE is at the valley bottom (γ_{Ig}), which implies a metastable state. Therefore, the existence of a local nanotwined structure is also demonstrated by the principle of minimum energy.

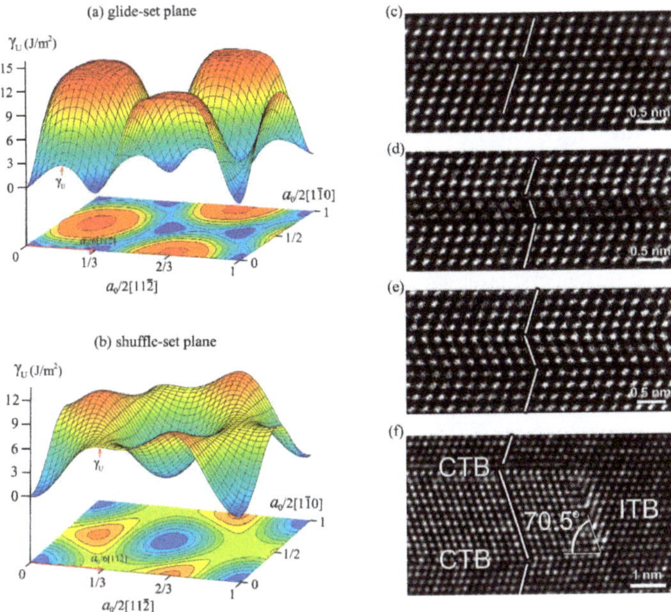

Figure 3. γ-surface and stacking fault region of cBN. (**a**) γ-surface of glide-set plane and (**b**) γ-surface of shuffle-set plane, with γ_U denoting energy barrier, red arrows indicating lowest energy path, respectively, (**c–f**) stacking fault region in cBN observed in experiment. Reproduced with permission from [47], copyright AIP Publishing, 2016.

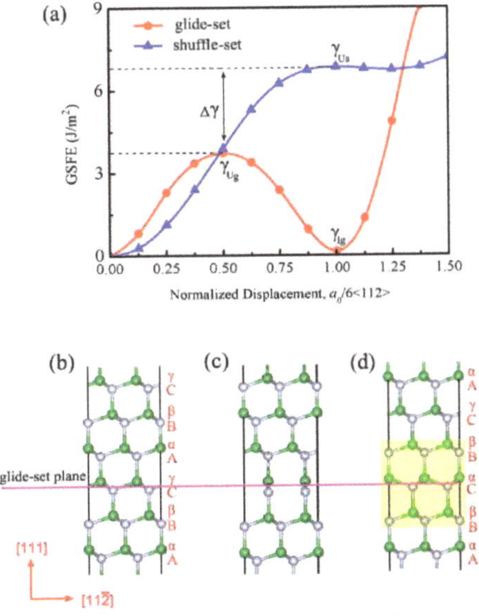

Figure 4. (**a**) Calculated GSFE curves for cBN in (111)<112> slip system, where γ_{Ug} and γ_{Us} denote energy barrier on glide-set and shuffle-set planes, respectively. Atomic configurations corresponding to three key structures in (111)<112> slip system: (**b**) initial configuration, (**c**) unstable configuration, and (**d**) metastable configuration, where light-yellow regions correspond to the stacking fault region.

3.2. Detwinning Mechanism of nt-cBN

Figure 5 shows the shear stress-strain (σ-ε) and energy-strain (E-ε) curves of nt-cBN during shear straining along the $(111)[11\bar{2}]$ direction, where both the curves exhibit a distinct zigzag feature. This zigzag feature is significantly different from cBN, as reported previously [48], fails directly during the shear straining due to graphitization. It can be seen in Figure 5 that σ and E increase quickly as the applied shear strain increases from $\varepsilon_0 = 0$ to $\varepsilon_1 = 0.23$ when the stress reaches 62.1 GPa, which is slightly smaller than the stress (67.2 GPa) before the graphitization of cBN subjected to shear deformation along the easy shear direction. When the strain increases from $\varepsilon_1 = 0.23$ to $\varepsilon_2 = 0.24$, the stress and total energy fall sharply. In order to clarify the cause of the zigzag manner, we extract the atomic configurations at some key points, as shown in Figure 6. It can be found that the root cause of this phenomenon is atomic reconfiguration induced by partial slip on the glide-set plane, in which the old chemical bond B_1-N_1-B_2 is broken, and a new chemical bond B_2-N_1-B_3 is formed, as shown in Figure 6a,b.

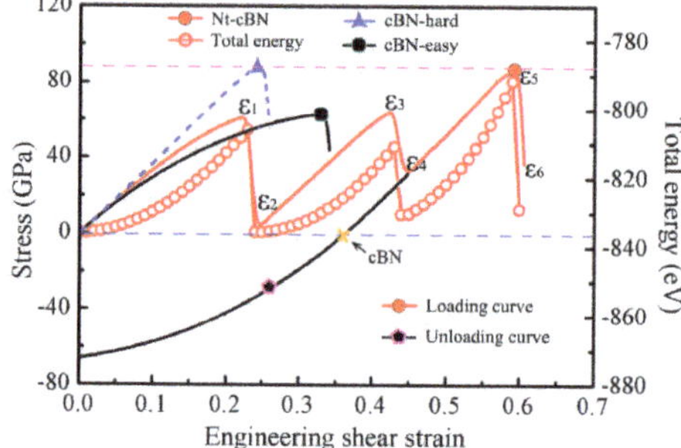

Figure 5. Calculated σ-ε and E-ε curves of nt-cBN subjected to shear straining and unloading, σ-ε curves of cBN sheared along $(111)[11\bar{2}]$ easy shear direction and $(111)[\bar{1}\bar{1}2]$ hard shear direction also provided for comparison.

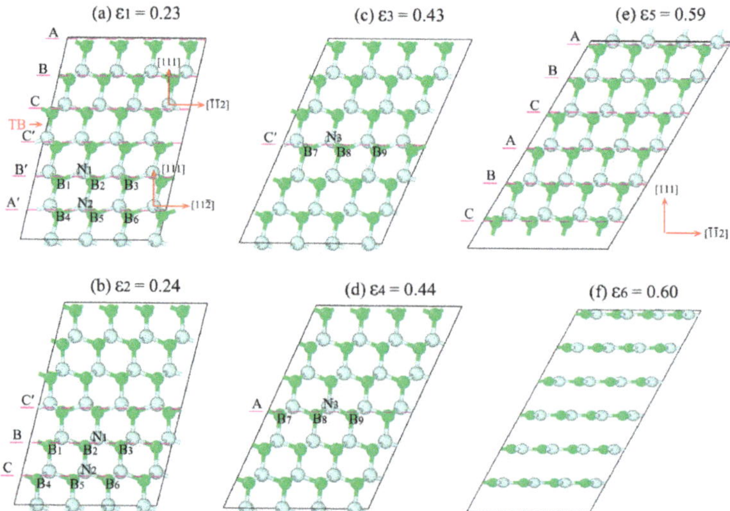

Figure 6. (**a**) to (**f**) are typical atomic configurations of nt-cBN sheared along $(111)[11\bar{2}]$ easy shear direction during $0.23 \leq \varepsilon \leq 0.60$, respectively, with upper and lower halves corresponding to easy- and hard-shear directions in (**a**).

The formation of the new metastable structures (Figure 6b) can further resist deformation induced by the further shear strain. From $\varepsilon_2 = 0.24$ to $\varepsilon_3 = 0.43$, σ increases from the valley of 3.7 GPa to the peak of 65.8 GPa. Then, the same atomic reconfiguration occurs in the layer C′, resulting in the break of the chemical bond B_7-N_3-B_8, and the formation of chemical bond B_8-N_3-B_9, as shown in Figure 6c,d. This process will be repeated during the shear straining until all the atomic layers in the easy shear direction are transformed into that in the hard shear direction, and, correspondingly, the sequence of the layers would change from ...ABC|C′B′A′... [Figure 6a] to ABCABC... [Figure 6e] when the detwinning of nt-cBN by partial slip is finished. The further increase of strain would induce the deformation along the hard shear direction of the cBN, as shown in Figure 6e, and at $\varepsilon_5 = 0.59$ the ultimate stress reaches 87.2 GPa, which is almost equal to the graphitization stress (87.4 GPa) of cBN along the hard shear direction, as shown in Figure 5. Lattice instability occurs as $\varepsilon \geq \varepsilon_6 = 0.60$, leading to graphite-like layered structures, as shown in Figure 6f.

The above process describes accurately the one-to-one correspondence between σ (or E) and atomic reconfiguration. In general, such kind of detwinning mechanism can lead to an unprecedented increase in intragranular deformation resistance for nt-cBN. More detailed information on the atomic reconfiguration during the deformation of cBN and nt-cBN are shown in Supplementary Movies S5–S7. Supplementary Movies S5 and S6 show the structural changes along the easy and hard shear directions of cBN subjected to shear deformations, respectively, where one can see that cBN fails directly by graphitization, which is consistent with the previous report [47]. By contrast, the detwinning mechanism of nt-cBN by partial slip on the glide-set plane is shown in Supplementary Movie S7, where the continuous partial slip can effectively delay the structural graphitization, contributing to the remarkable increases of the maximum strain and intragranular deformation resistance.

To verify the fascinating phase transition from nt-cBN to cBN, we unload the stress from $\varepsilon_4 = 0.44$ by decreasing the strain until $\sigma = 0$, as shown in Figure 5. After unloading, the lattice vectors of the cell are along the $[\bar{1}12]$, $[1\bar{1}0]$ and $[11\bar{2}]$ directions, which are exactly identical with those of the perfect crystal cBN cell, indicating the nt-cBN has been completely detwinned. Figure 7 shows the comparison between the atomic structure of relaxed cBN at zero stress and that of detwinned nt-cBN after unloading, where there are two findings. First, the positions of the atoms in the two structures overlap completely, second, the angle between lattice vectors $[\bar{1}12]$ and $[11\bar{2}]$ is approximately 70.55°, which is consistent with the experimental observations [47], as shown in Figure 3f.

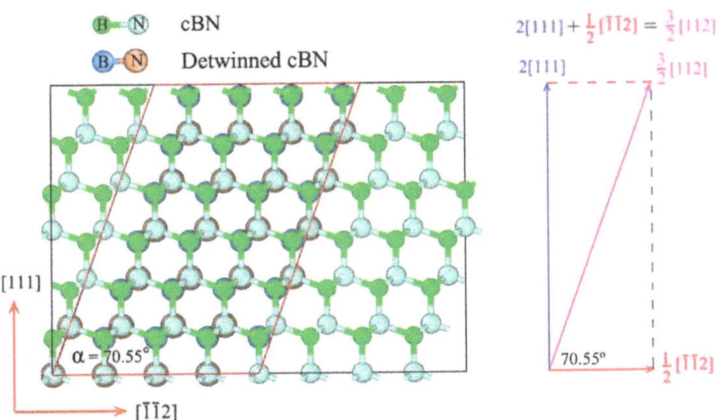

Figure 7. Comparison between atomic structure of relaxed cBN at zero stress and that of detwinned cBN after unloading.

4. Summary

We reported a detwinning mechanism for nt-cBN, which may result in extremely high hardness, which can describe accurately the one-to-one correspondence between the stress or the total energy and the atomic reconfiguration of nt-cBN subjected to shear straining. The atomic layers in the easy shear direction can be shifted into the hard shear direction by partial slip, which also leads to the decrease of the stress and total energy of the system and formation a new stable structure. This characteristic should be conducive to the stability of the structure undergoing lager shear strain. Moreover, the twin boundary energies of nt-diamond and nt-cBN are very low, which indicates that the twin structure is as stable as its single crystal counterparts. The excellent thermal stability of nt-diamond and nt-cBN has also been systematically studied by ab initio molecular dynamics (MD) simulations and a stabilization mechanism for nanotwin was revealed. These results can not only account for the unprecedented hardness and excellent stability of nanotwinned structure but offer novel insights into the deformation-induced structural transformation as well. It can also extend our understanding of the deformation mechanism of nanostructured strong covalent materials, which would be useful in guiding the design of ultrahard materials.

Supplementary Materials: The following are available online at http://www.mdpi.com/2079-4991/9/8/1117/s1, Movie S1. Ab initio molecular dynamics simulations under T = 1500 K for diamond in front view. Movie S2. Ab initio molecular dynamics simulations under T = 1500 K for nt-diamond in front view. Movie S3. Ab initio molecular dynamics simulations under T = 2000 K for cBN in front view. Movie S4. Ab initio molecular dynamics simulations under T = 2000 K for nt-cBN in front view. Movie S5. Single crystal cBN undergoes lattice instability along easy $(111)[11\bar{2}]$ shear direction, resulting in well-known graphitization. Movie S6. Single crystal cBN undergoes lattice instability along hard $(111)[\bar{1}\bar{1}2]$ shear direction, resulting in well-known graphitization. Movie S7. Typical phase transition from twin (nt-cBN) to cube (cBN) by partial slip on the glide-set plane, and further loading along the hard $(111)[\bar{1}\bar{1}2]$ shear direction may eventually results in structural graphitization.

Author Contributions: B.Y., X.P. and T.F. conceived and designed the simulations; B.Y. performed the simulations; B.Y., S.S., C.H., X.C. and D.Y. analyzed the data; B.Y., X.P. and T.F. wrote the paper.

Funding: This research was funded by the Fundamental Research Funds for the Central Universities (grant no. 2018CDYJSY0055), National Natural Science Foundation of China (grant no. 11332013, 11802045, and 11802047), National Postdoctoral Program for Innovative Talents (Grant No. BX20190039), the Postdoctoral Program for Innovative Talents of Chongqing (grant no. CQBX201804), and the China Postdoctoral Science Foundation funded project (grant no. 2018M631058).

Acknowledgments: The authors gratefully acknowledge the financial supports, and First-principles calculations were carried out at Supercomputing Center of Lv Liang Cloud Computing Center in China.

Conflicts of Interest: The authors declare no competing financial interest.

References

1. Lu, C.; Li, Q.; Ma, Y.; Chen, C. Extraordinary indentation strain stiffening produces superhard tungsten nitrides. *Phys. Rev. Lett.* **2017**, *119*, 115503. [CrossRef] [PubMed]

2. Huang, C.; Peng, X.; Fu, T.; Zhao, Y.; Feng, C.; Lin, Z.; Li, Q. Nanoindentation of ultra-hard cbn films: A molecular dynamics study. *Appl. Surf. Sci.* **2017**, *392*, 215–224. [CrossRef]

3. An, Q.; Goddard, W.A., 3rd. Atomistic origin of brittle failure of boron carbide from large-scale reactive dynamics simulations: Suggestions toward improved ductility. *Phys. Rev. Lett.* **2015**, *115*, 105501. [CrossRef] [PubMed]

4. Brazhkin, V.V.; Lyapin, A.G.; Hemley, R.J. Harder than diamond: Dreams and reality. *Philos. Mag. A* **2002**, *82*, 231–253. [CrossRef]

5. Huang, C.; Peng, X.; Yang, B.; Xiang, H.; Sun, S.; Chen, X.; Li, Q.; Yin, D.; Fu, T. Anisotropy effects in diamond under nanoindentation. *Carbon* **2018**, *132*, 606–615. [CrossRef]

6. Huang, C.; Peng, X.; Yang, B.; Chen, X.; Li, Q.; Yin, D.; Fu, T. Effects of strain rate and annealing temperature on tensile properties of nanocrystalline diamond. *Carbon* **2018**, *136*, 320–328. [CrossRef]

7. Yang, B.; Peng, X.; Huang, C.; Yin, D.; Xiang, H.; Fu, T. Higher strength and ductility than diamond: Nanotwinned diamond/cubic boron nitride multilayer. *ACS Appl. Mater. Int.* **2018**, *10*, 42804–42811. [CrossRef]

8. Yang, B.; Peng, X.; Huang, C.; Wang, Z.; Yin, D.; Fu, T. Strengthening and toughening by partial slip in nanotwinned diamond. *Carbon* **2019**, *150*, 1–7. [CrossRef]

9. Tanigaki, K.; Ogi, H.; Sumiya, H.; Kusakabe, K.; Nakamura, N.; Hirao, M.; Ledbetter, H. Observation of higher stiffness in nanopolycrystal diamond than monocrystal diamond. *Nat. Commun.* **2013**, *4*, 2343. [CrossRef]

10. Mochalin, V.N.; Shenderova, O.; Ho, D.; Gogotsi, Y. The properties and applications of nanodiamonds. *Nat. Nanotechnol.* **2011**, *7*, 11–23. [CrossRef]

11. Lu, K.; Lu, L.; Suresh, S. Strengthening materials by engineering coherent internal boundaries at the nanoscale. *Science* **2009**, *324*, 349–352. [CrossRef]

12. Wang, Z.; Saito, M.; McKenna, K.P.; Gu, L.; Tsukimoto, S.; Shluger, A.L.; Ikuhara, Y. Atom-resolved imaging of ordered defect superstructures at individual grain boundaries. *Nature* **2011**, *479*, 380–383. [CrossRef]

13. Wang, Z.; Saito, M.; McKenna, K.P.; Ikuhara, Y. Polymorphism of dislocation core structures at the atomic scale. *Nat. Commun.* **2014**, *5*, 3239. [CrossRef]

14. Hall, E.O. The deformation and aging of mild steel. *Proc. Phys. Soc. Lond. Ser. B* **1951**, *64*, 495–502.

15. Petch, N.J. The cleavage strength of polycrystals. *J. Iron Steel Inst.* **1953**, *174*, 25–28.

16. Fu, T.; Peng, X.; Huang, C.; Weng, S.; Zhao, Y.; Wang, Z.; Hu, N. Strain rate dependence of tension and compression behavior in nano-polycrystalline vanadium nitride. *Ceram. Int.* **2017**, *43*, 11635–11641. [CrossRef]

17. Yip, S. The strongest size. *Nature* **1998**, *86*, 713–720. [CrossRef]

18. Sumiya, H.; Irifune, T. Hardness and deformation microstructures of nano-polycrystalline diamonds synthesized from various carbons under high pressure and high temperature. *J. Mater. Res.* **2011**, *22*, 2345–2351. [CrossRef]

19. Irifune, T.; Kurio, A.; Sakamoto, S.; Inoue, T.; Sumiya, H. Materials: Ultrahard polycrystalline diamond from graphite. *Nature* **2003**, *421*, 599–600. [CrossRef]

20. Solozhenko, V.L.; Kurakevych, O.O.; Le Godec, Y. Creation of nanostuctures by extreme conditions: High-pressure synthesis of ultrahard nanocrystalline cubic boron nitride. *Adv. Mater.* **2012**, *24*, 1540–1544. [CrossRef]

21. Dubrovinskaia, N.; Solozhenko, V.L.; Miyajima, N.; Dmitriev, V.; Kurakevych, O.O.; Dubrovinsky, L. Superhard nanocomposite of dense polymorphs of boron nitride: Noncarbon material has reached diamond hardness. *Appl. Phys. Lett.* **2007**, *90*, 101912. [CrossRef]

22. Hall, E.O. The Deformation and Ageing of Mild Steel. Discussion of Results. *Proc. Phys. Soc. Lond. Sect. B* **1951**, *64*, 747. [CrossRef]

23. Tian, Y.; Xu, B.; Yu, D.; Ma, Y.; Wang, Y.; Jiang, Y.; Hu, W.; Tang, C.; Gao, Y.; Luo, K.; et al. Ultrahard nanotwinned cubic boron nitride. *Nature* **2013**, *493*, 385–388. [CrossRef]

24. Lu, K. Stabilizing nanostructures in metals using grain and twin boundary architectures. *Nat. Rev. Mater.* **2016**, *1*, 16019. [CrossRef]

25. Huang, Q.; Yu, D.L.; Xu, B.; Hu, W.T.; Ma, Y.M.; Wang, Y.B.; Zhao, Z.S.; Wen, B.; He, J.L.; Liu, Z.Y.; et al. Nanotwinned diamond with unprecedented hardness and stability. *Nature* **2014**, *510*, 250. [CrossRef]

26. Cheng, Z.; Zhou, H.; Lu, Q.; Gao, H.; Lu, L. Extra strengthening and work hardening in gradient nanotwinned metals. *Science* **2018**, *362*, 559. [CrossRef]

27. Zhou, X.; Li, X.Y.; Lu, K. Enhanced thermal stability of nanograined metals below a critical grain size. *Science* **2018**, *360*, 526–530. [CrossRef]

28. Pan, Q.; Zhou, H.; Lu, Q.; Gao, H.; Lu, L. History-independent cyclic response of nanotwinned metals. *Nature* **2017**, *551*, 214–217. [CrossRef]

29. You, Z.; Lu, L. Deformation and fracture mechanisms of nanotwinned metals. *Natl. Sci. Rev.* **2017**, *4*, 519–521. [CrossRef]

30. Li, X.; Yin, S.; Oh, S.H.; Gao, H. Hardening and toughening mechanisms in nanotwinned ceramics. *Scr. Mater.* **2017**, *133*, 105–112. [CrossRef]

31. Taheri Mousavi, S.M.; Zou, G.; Zhou, H.; Gao, H. Anisotropy governs strain stiffening in nanotwinned-materials. *Nat. Commun.* **2018**, *9*, 1586. [CrossRef]

32. Li, B.; Sun, H.; Chen, C. Reply to 'anisotropy governs strain stiffening in nanotwinned-materials'. *Nat. Commun.* **2018**, *9*, 1585. [CrossRef]

33. Perdew, J.P.; Burke, K.; Ernzerhof, M. Generalized gradient approximation made simple. *Phys. Rev. Lett.* **1996**, *77*, 3865–3868. [CrossRef]

34. Kresse, G.; Furthmuller, J. Efficient iterative schemes for ab initio total-energy calculations using a plane-wave basis set. *Phys. Rev. B* **1996**, *54*, 11169–11186. [CrossRef]

35. Kresse, G.; Joubert, D. From ultrasoft pseudopotentials to the projector augmented-wave method. *Phys. Rev. B* **1999**, *59*, 1758–1775. [CrossRef]

36. Monkhorst, H.J.; Pack, J.D. Special points for brillouin-zone integrations. *Phys. Rev. B* **1976**, *13*, 5188–5192. [CrossRef]

37. Guan, J.; Zhu, Z.; Tomanek, D. Phase coexistence and metal-insulator transition in few-layer phosphorene: A computational study. *Phys. Rev. Lett.* **2014**, *113*, 046804. [CrossRef]

38. Yang, B.; Peng, X.; Xiang, H.; Yin, D.; Huang, C.; Sun, S.; Fu, T. Generalized stacking fault energies and ideal strengths of MC systems (M = Ti, Zr, Hf) doped with si/al using first principles calculations. *J. Alloys Compd.* **2018**, *739*, 431–438. [CrossRef]

39. Roundy, D.; Krenn, C.R.; Cohen, M.L.; Morris, J.W., Jr. Ideal shear strengths of fcc aluminum and copper. *Phys. Rev. Lett.* **1999**, *82*, 2713–2716. [CrossRef]

40. Roundy, D.; Krenn, C.R.; Cohen, M.L.; Morris, J.W. The ideal strength of tungsten. *Philos. Mag. A* **2001**, *81*, 1725–1747. [CrossRef]

41. Rodney, D.; Ventelon, L.; Clouet, E.; Pizzagalli, L.; Willaime, F. Ab initio modeling of dislocation core properties in metals and semiconductors. *Acta Mater.* **2017**, *124*, 633–659. [CrossRef]

42. Xiang, H.; Li, H.; Fu, T.; Huang, C.; Peng, X. Formation of prismatic loops in aln and gan under nanoindentation. *Acta Mater.* **2017**, *138*, 131–139. [CrossRef]

43. Zhang, M.; Liu, H.; Li, Q.; Gao, B.; Wang, Y.; Li, H.; Chen, C.; Ma, Y. Superhard BC_3 in cubic diamond structure. *Phys. Rev. Lett.* **2015**, *114*, 015502. [CrossRef]

44. Wang, P.; He, D.; Wang, L.; Kou, Z.; Li, Y.; Xiong, L.; Hu, Q.; Xu, C.; Lei, L.; Wang, Q.; et al. Diamond-cbn alloy: A universal cutting material. *Appl. Phys. Lett.* **2015**, *107*, 101901. [CrossRef]

45. Zhang, Y.; Sun, H.; Chen, C. Atomistic deformation modes in strong covalent solids. *Phys. Rev. Lett.* **2005**, *94*, 145505. [CrossRef]

46. Telling, R.H.; Pickard, C.J.; Payne, M.C.; Field, J.E. Theoretical strength and cleavage of diamond. *Phys. Rev. Lett.* **2000**, *84*, 5160–5163. [CrossRef]

47. Zheng, S.; Zhang, R.; Huang, R.; Taniguchi, T.; Ma, X.; Ikuhara, Y.; Beyerlein, I.J. Structure and energetics of nanotwins in cubic boron nitrides. *Appl. Phys. Lett.* **2016**, *109*, 081901. [CrossRef]

48. Zhang, R.F.; Veprek, S.; Argon, A.S. Anisotropic ideal strengths and chemical bonding of wurtzite bn in comparison to zincblende bn. *Phys. Rev. B* **2008**, *77*, 172103. [CrossRef]

MDPI

St. Alban-Anlage 66

4052 Basel

Switzerland

Tel. +41 61 683 77 34

Fax +41 61 302 89 18

www.mdpi.com

Nanomaterials Editorial Office

E-mail: nanomaterials@mdpi.com

www.mdpi.com/journal/nanomaterials